UNDERSTANDING MEMS

UNDERSTANDING MEMS
PRINCIPLES AND APPLICATIONS

Luis Castañer

Department of Electronic Engineering
Universitat Politécnica de Cataluña, Spain

Library of Congress Cataloging-in-Publication Data

Castañer, Luis, author.
   Understanding MEMS : principles and applications / Luis Castañer.
      pages cm
   Includes bibliographical references and index.
   ISBN 978-1-119-05542-6 (cloth)
   1. Microelectromechanical systems.   I. Title.
   TK7875.C37 2015
   621.381–dc23

                                        2015024280

A catalogue record for this book is available from the British Library.

ISBN: 9781119055426

Set in 10/12pt Times by Aptara Inc., New Delhi, India

Printed in Singapore by C.O.S. Printers Pte Ltd

1   2016

To my wife Pamen, daughters Maya and Olga, grandsons Teo, Raul and Marc-Eric and granddaughter Claudia

Contents

Preface

The field of microelectromechanical systems (MEMS) has greatly expanded in recent years since Richard Feyman's speech – 'there is plenty of room at the bottom' – opened the minds of researchers and companies to the possibility of exploring the potential of microstructures of minute dimensions.

The need for skilled professionals has steadily grown as businesses and research labs engage in challenging projects. Students are motivated because they are aware that they have, in their phones, watches or tablets, accelerometers, compasses and sensors and useful applications relying on them. Students see in the MEMS field, job opportunities and exciting career prospects.

Ultimately those devices have to work together with front-end electronics and digital processing to interface with the user. Students in electrical engineering and computer science departments in universities around the world are probably the most exposed to the field. Simultaneously, engineers and professionals working in the electronics and semiconductor industry face the challenge of integrating MEMS devices into chips, systems on chips or, broadly speaking, into electronic systems. The added value of those devices enables the expansion of high tech businesses.

In my experience of teaching MEMS in an electronic engineering department to engineering students from many countries I have faced the difficulty of selecting material for one-semester course and having to decide on the depth and breadth of the subjects covered.

The field is inherently multidisciplinary, and if the basics are not sufficiently covered, students will not achieve the intellectual satisfaction of a full understanding. However, if the coverage is too complex it cannot be extended to the various fundamental domains underlying the field. Solving problems is an important part of the learning process as it allows for concepts to be reviewed.

Those are the reasons why I have chosen to approach the subject using analytical solutions as far as possible, but with the help of two software tools: one very popular among science and engineering students, Matlab; and the other, very popular among electrical engineering students, PSpice. I have used Matlab to solve ordinary differential equations subject to boundary or initial conditions, applied to mechanical, electromechanical, electrokinetic and thermal problems. This allows numerical results to be found quickly which can then be discussed and put into context.

I have used PSpice to solve Laplace transforms of transfer functions and to solve electrical equivalent circuits of lumped thermal problems. Apart from the clarity of analytical solutions, this approach places the subject of MEMS in the same tool environment as other subjects the student will already have taken. Commonality of tools is important at this level of the learning process, because it means that students do not have to spend a significant amount of time learning how to use new software.

The book includes 52 worked examples in the text and 100 solved problems in the appendices, organized by chapters. In my view this allows this textbook to be used not only as support material for a conventional course but also as a self-study resource for distance learning. I am very greatful to faculty colleagues, researchers and students with whom I have interacted all these years that have taken me to complete this book.

Luis Castañer

April 2015

Barcelona, Spain

About the Companion Website

This book is accompanied by a companion website:

www.wiley.com/go/castaner/understandingmems

The website includes:

- Matlab
- PSpice codes
- Chapter viewgraphs

1

Scaling of Forces

There are a number of important forces in the field of microelectromechanical systems (MEMS). However, their relative importance does not necessarily match the importance they have in the macroworld. This chapter is concerned with the scaling of these forces to small dimensions. Weight, elastic, electrostatic, capillary, piezoelectric, magnetic and dielectrophoretic forces are examined and a scaling factor identified for all of them.

1.1 Scaling of Forces Model

The integration of complex and powerful systems in silicon for a large variety of applications stems from the miniaturization of electronic devices and components. Electromechanical components that were bulky, heavy and inefficient can now be miniaturized using MEMS technology. Here, mechanical moving parts are used both for sensing devices and actuators. The main forces present in the operation of these components depend on the geometrical dimensions, and thereby, when the dimensions are scaled down, the magnitudes of these forces change, creating a different scenario compared to the macroworld.

Given a force F that depends on a number of geometrical dimensions a_i and on a number of parameters γ_j, we have

$$F = F(a_i, \gamma_j). \tag{1.1}$$

When all dimensions are scaled by the same factor α, the force changes to

$$F_\alpha = F(\alpha a_i, \gamma_j), \tag{1.2}$$

provided all the parameters γ_j do not depend on the geometrical dimensions. The ratio of the forces before and after the dimension scaling is given by

$$\frac{F_\alpha}{F} = \frac{F(\alpha a_i, \gamma_j)}{F(a_i, \gamma_j)}. \tag{1.3}$$

Understanding MEMS: Principles and Applications, First Edition. Luis Castañer.
© 2016 John Wiley & Sons, Ltd. Published 2016 by John Wiley & Sons, Ltd.
Companion Website: www.wiley.com/go/castaner/understandingmems

Generally, when analytical models are used in simplified cases, the result of equation (1.3) provides a direct relation to a power n of the scaling factor α,

$$\frac{F_\alpha}{F} = \alpha^n, \tag{1.4}$$

meaning that when the dimensions are scaled down by a factor α, the force scales down by a factor α^n.

1.2   Weight

As our first application of the rule provided by equation (1.3) we consider the scaling of weight. Imagine that we have a body of length L, width W and thickness t. The weight of this body is given by

$$F = \rho_m g L W t, \tag{1.5}$$

where ρ_m is the material density and g the acceleration due to gravity. If all dimensions are scaled by a factor α, the length becomes αL, the width becomes αW and the thickness becomes αt, and so the scaled weight is

$$F_\alpha = \rho_m g \alpha L \alpha W \alpha t = \alpha^3 \rho_m g L W t, \tag{1.6}$$

and the ratio of forces after and before scaling is given by

$$\frac{F_\alpha}{F} = \alpha^3. \tag{1.7}$$

Equation (1.7) tells us that the weight scales down as the third power of the scaling factor, so if we reduce all dimensions by a factor of 10 ($\alpha = 0.1$), the weight is multiplied by a factor of $\alpha^3 = 0.001$).

It will become clear in the next sections that when electromechanical structures are miniaturized, the weight loses the importance it has in the macroworld and other forces become the main players.

1.2.1   Example: MEMS Accelerometer

A MEMS accelerometer has an inertial mass made up of a plate of silicon bulk material of 500 μm side and 500 μm thickness. Calculate the force developed when subject to an acceleration ten times that due to gravity.

Taking into account that the density of silicon is $2329\,\text{kg/m}^3$ and that the volume is $500 \times 500 \times 500 \times 10^{-18}\,\text{m}^3$, the force is given by

$$F = 2.85 \times 10^{-5}\,\text{N}.$$

If all dimensions are reduced by a factor of 10 ($\alpha = 0.1$) the weight reduces to $F_\alpha = 2.85 \times 10^{-8}\,\text{N}$.

1.3 Elastic Force

A body is deformed when it is subject to an external force. In equilibrium, the elastic force is the restoring force that compensates the external force. If the deformation is elastic, the initial dimensions of the body are recovered after the external force disappears. In a one-dimensional geometry and according to Hooke's law [1], the elastic force, F, is proportional to the deformation length δ, collinear with the force,

$$F = k\delta, \tag{1.8}$$

where k is the stiffness constant that is not independent of the geometry as will be shown in Chapter 3; for example, for a cantilever of rectangular cross-section with length L, width W and thickness t, subject to a force applied at the tip (see Figure 1.1), the stiffness is given by

$$k = \frac{EWt^3}{4L^3}, \tag{1.9}$$

where E is Young's elasticity modulus. We now proceed as in Section 1.1 and calculate the forces F and F_α before and after scaling:

$$F = \frac{EWt^3}{4L^3}\delta, \quad F_\alpha = \frac{E\alpha W \alpha^3 t^3}{4\alpha^3 L^3}\alpha\delta = \alpha^2 \frac{EWt^3}{4L^3}\delta. \tag{1.10}$$

The ratio between these two quantities is therefore

$$\frac{F_\alpha}{F} = \alpha^2. \tag{1.11}$$

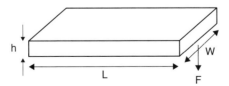

Figure 1.1   Geometry of a cantilever loaded at the tip

1.3.1   Example: AFM Cantilever

In atomic force microscopy tiny cantilevers with a very sharp tip are used to detect the force.
The cantilever acts as a soft spring. Calculate the force that will deflect the cantilever by 1 μm
for L = 200 μm, *W = 5* μm *and h = 2* μm.

By equation (1.9), $k = 0.081$ N/m, and by equation (1.8),

$$F = 8.1 \times 10^{-8} \text{N}.$$

Applying a dimension scaling with $\alpha = 0.1$, the force reduces to $F_\alpha = 8.1 \times 10^{-10}$ N.

1.4   Electrostatic Force

The electrostatic force between two plates is due to the electric field, **E**, that builds up when an
electric potential V is applied between them.[1] This is a very common way to make mechanical
parts move in today's microelectromechanical devices.

If we consider one of the two plates charged with a charge density σ, as shown in Figure 1.2(a), Gauss's law [2] allows to calculate the electric field created by the charged sheet as

$$\oint \overrightarrow{DdS} = \int \sigma dS = Q. \tag{1.12}$$

Signs in equation (1.12) are taken as positive for an electric field directed outward from the
differential volume, and $\overrightarrow{dS}$ is taken positive also directed outward from the face. As the
electric field is normal to the charged surface, only integrals extending over the top and bottom
surfaces of the volume are different from zero, so that

$$\int_{\text{top}} \epsilon \mathbf{E} dS + \int_{\text{bottom}} \epsilon \mathbf{E} dS = Q, \tag{1.13}$$

$$\epsilon \mathbf{E} A + \epsilon \mathbf{E} A = Q, \tag{1.14}$$

where A is the area of the surface. Then

$$\mathbf{E} = \frac{Q}{2\epsilon A}. \tag{1.15}$$

The Coulomb force that such a field exerts on the parallel plate with a charge of $-Q$ and at a
distance g is

$$F = -Q\mathbf{E} = -\frac{Q^2}{2\epsilon A}. \tag{1.16}$$

---

[1] We denote the electric field by **E** to distinguish it from the Young's modulus E.

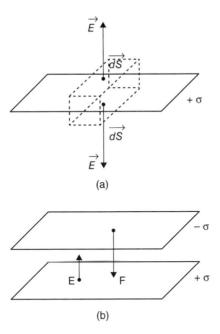

Figure 1.2 (a) Gauss's law for a sheet of charge σ, and (b) electric field and Coulomb force exerted on the upper plate

Since $Q = CV$ and $C = \epsilon A/g$,

$$F = \frac{\epsilon A V^2}{2g^2}. \tag{1.17}$$

As can be seen the force is downwards, that is to say, it is attractive between the plates and does not depend on the sign of the applied voltage as it is squared in the force equation (1.17). When we apply the scaling method we find that

$$F_\alpha = \frac{\epsilon A \alpha^2 V^2}{2g^2 \alpha^2}. \tag{1.18}$$

In equation (1.18), A is the area of the plates which scales as α^2, and g is the value of the gap between plates. The scaling factor of the force is

$$\frac{F_\alpha}{F} = \alpha^0 = 1. \tag{1.19}$$

This is a very important result showing that the electrostatic force is independent of the scaling factor and can be very high compared to other forces in the microworld. However, it can be correctly argued that reducing the distance between plates increases the electric field and the devices may be damaged by breakdown. To prevent this situation, we can consider a different scaling scenario in which the value of the electric field is kept constant. As the electric field is $\mathbf{E} = V/g$, equation (1.18) can be written as

$$F = \frac{\epsilon AV^2}{2g^2} = \frac{\epsilon A\mathbf{E}^2 g^2}{2g^2} = \frac{\epsilon A\mathbf{E}^2}{2}, \quad F_\alpha = \frac{\epsilon A\alpha^2 \mathbf{E}^2}{2} \tag{1.20}$$

and hence

$$\frac{F_\alpha}{F} = \alpha^2. \tag{1.21}$$

Here we see that in this scenario the scaling follows an α^2 rule instead.

1.4.1   Example: MEMS RF Switch

In a MEMS RF switch two metal plates $250 \times 250 \ \mu m^2$ *are driven by a voltage of 9 V. Calculate the force required to close the 5 μm gap between them.*

If we suppose that between the two plates there is air, the permittivity is $\epsilon = 8.85 \times 10^{-12}$ F/m, and the force can be calculated from equation (1.18):

$$F = \frac{\epsilon AV^2}{2g^2} = 8.85 \times 10^{-12} \frac{250 \times 10^{-6} \times 250 \times 10^{-6} \times 9^2}{2 \times 5^2 \times 10^{-12}} = 8.96 \times 10^{-7} \ \text{N}.$$

If the dimensions are scaled by a factor of $\alpha = 0.1$ the force remains equal if equation (1.19) applies or 8.96×10^{-9} N if equation (1.21) applies.

1.5   Capillary Force

On the surface of a liquid the molecules are attracted by the other molecules inside the volume but do not have the attraction from the surroundings above the surface. This creates a situation where the molecules rearrange in order to expose the minimum surface. If an observer wants to increase the surface exposed to the ambient, he necessarily has to do some work. This work, dW, is proportional to the increase in area, dA [3]:

$$dW = \gamma dA. \tag{1.22}$$

The proportionality constant γ is the surface tension and has units of J/m^2 or, equivalently, N/m. Thus the surface tension is a measure of the surface energy per unit area.

When a liquid drop is in equilibrium, there is a pressure increase ΔP inside the drop, known as Laplace pressure, to prevent collapse. ΔP is related to the surface tension by

$$\Delta P = \gamma C, \tag{1.23}$$

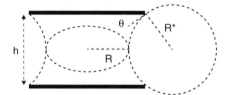

Figure 1.3 Parallel plates with droplet of liquid in between

where C is the curvature of the drop given by

$$C = \frac{1}{R} + \frac{1}{R^*},\qquad(1.24)$$

in which R and R^* are the radius of two mutually orthogonal circles drawn at a tangency point of the drop surface. R is the radius of the circle that lies inside the drop, and R^* that of the one lying outside and takes negative sign. This is shown in Figure 1.3, where the example of two parallel plates having a drop of liquid trapped inside is considered.

Due to equilibrium of surface tensions, a liquid on a substrate has a contact angle θ shown in Figure 1.3 (see also Section 7.5). We have that

$$\frac{h}{2} = R^* \cos\theta,\qquad(1.25)$$

hence

$$\Delta P = \gamma \left(\frac{1}{R} - \frac{2\cos\theta}{h} \right).\qquad(1.26)$$

In many MEMS applications $R \gg h$, and then equation (1.26) simplifies to

$$\Delta P \simeq -\gamma \frac{2\cos\theta}{h}.\qquad(1.27)$$

Equation (1.27) shows that if the contact angle $0 < \theta < \pi/2$, then $\Delta P < 0$ and the force is inwards with respect to the liquid, whereas for $\pi/2 < \theta < \pi$, $\Delta P > 0$ and the force is outwards. If we suppose that the plates shown in Figure 1.3 are circular with radius R, then the force developed by capillarity between the two plates is

$$F = \pi R^2 \Delta P = -\gamma \frac{2\pi \cos\theta R^2}{h}.\qquad(1.28)$$

When the geometrical dimensions are scaled,

$$F_\alpha = \gamma \frac{2\pi \cos\theta \alpha^2 R^2}{\alpha g},\qquad(1.29)$$

and the ratio is given by

$$\frac{F_\alpha}{F} = \alpha^1.\qquad(1.30)$$

1.5.1 Example: Wet Etching Force

We have a MEMS process involving wet etching of a sacrificial layer between two parallel circular plates of radius $R = 2500$ μm. If the gap between the plates is $g = 2$ μm, the surface tension is $\gamma = 72.9 \times 10^{-3}$ N/m and the contact angle between the liquid and the substrate is 70°, calculate the force between the plates.

As the contact angle is smaller than $\pi/2$, the force is attractive and the value can be calculated from equation (1.28):

$$F = \gamma \frac{2\pi \cos \theta R^2}{g} = 72.9 \times 10^{-3} \frac{2\pi \cos 70 (2500 \times 10^{-6})^2}{2 \times 10^{-6}} = 0.48 \, \text{N}.$$

1.6 Piezoelectric Force

Piezoelectricity is a property of some materials that generate electric charge when mechanically stressed and undergo a deformation when biased by an electric field. This phenomenon arises from a change in the crystallization of a material when subject to a process of simultaneous application of a high electric field and a high temperature known as 'polling'. Electrical dipoles are generated during polling that remain in the material thereafter. Piezoelectric layers can be used as displacement actuators or as force generators against a restraint.

The main equation of the direct piezoelectric effect is

$$D = dT + \epsilon \mathbf{E}, \tag{1.31}$$

where D is the electrical displacement vector, d the piezoelectric coefficient, ϵ the permittivity of the material, $\mathbf{E}$ the electric field and T the mechanical stress. In equation (1.31) the piezoelectric effect is anisotropic (see Chapter 4) and d is a tensor.

It can be seen that the electrical displacement D has two components: one conventional due to the electric field applied, and the other due to the mechanical stress. Conversely, the inverse piezoelectric effect is described by the equation

$$S = sT + d^T \mathbf{E} \tag{1.32}$$

where S is the strain (or relative deformation), s the compliance and d^T the transpose of the piezoelectric coefficient tensor. When an electric field is applied, assuming that the material has a force restraint F working against deformation, equation (1.32) can be written as

$$S = -s\frac{F}{A} + d\mathbf{E}, \tag{1.33}$$

where A is the cross-section of the material. Equation (1.33) shows that in the absence of any restraint ($F = 0$), the maximum displacement, or maximum stroke, is $S_{\text{max}} = d\mathbf{E}$ and the

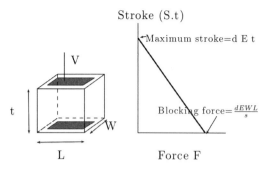

Figure 1.4   Force as a function of stroke

maximum stress $(F/A)_{max}$ happens for zero deformation $(S = 0)$ and is given, as shown in Figure 1.4. by

$$\left.\frac{F}{A}\right|_{max} = \frac{d}{s}\mathbf{E}.$$ (1.34)

If we consider, as an indicator for the scaling scenario, the maximum force value, or blocking force, and that the applied electric field is given by V/t, where t is the material thickness and V the applied voltage, the force scales as

$$F = \frac{dAV}{st}.$$ (1.35)

Applying the scaling model,

$$\frac{F_\alpha}{F} = \alpha^1,$$ (1.36)

and if the scaling is performed at constant electric field, then

$$\frac{F_\alpha}{F} = \alpha^2.$$ (1.37)

1.6.1   Example: Force in Film Embossing

We have a piezoelectric material 2 μm thick and 200 μm × 200 μm in area. We apply a voltage of 10 V across the film and we want to know the deformation in the direction of the electric field that is achieved. Calculate the value of the maximum force that can be put in a wall preventing the deformation of the material, such as occurs in film embossing. We know that the film is made of ZnO, $d = 12 \times 10^{-12}$ CN^{-1} and $s = 7 \times 10^{-12}$.

We first calculate the maximum strain S_{max},

$$S_{max} = d\mathbf{E} = d\frac{V}{t} = 6 \times 10^{-4}, \quad \Delta t = St = 1.2\,\text{nm},$$

and the maximum force (or blocking force) for zero deformation is given by

$$F = A\frac{V}{t}\frac{d}{s} = (200 \times 10^{-6})^2 \frac{100}{2 \times 10^{-6}} \frac{12 \times 10^{-12}}{7 \times 10^{-12}} = 0.34\,\text{N}.$$

As can be seen, the piezoelectric actuators can generate large forces but small displacements.

1.7 Magnetic Force

One important MEMS application is the measurement of magnetic field for compasses [4, 5]. One way to detect a magnetic field uses the Lorentz force that develops when a wire carrying an electric current intensity I is immersed in a magnetic flux density B. If we know the intensity value and the length of the wire L, the Lorentz force is given by

$$F = I\vec{L} \times \vec{B} \qquad (1.38)$$

where $\times$ indicates the cross product of the magnetic field vector and the wire length vector. This force is orthogonal to both the magnetic field and the wire direction. If the magnetic field and the wire are orthogonal, then the magnitude of the force is simply given by

$$F = ILB. \qquad (1.39)$$

When scaling equation (1.39), one has to take into account that decreasing dimensions, most of the time, require also reducing the cross-section of the wire. If the current I is constant in the scaling, then the current density will increase and the ohmic losses will also increase as the resistance of the wire increases. It is then useful to consider that the magnitude that is kept constant in the scaling is the current density $J = I/A$, where A is the cross-section of the wire. Hence, equation (1.39) can be written as

$$F = JALB, \quad F_\alpha = J\alpha^2 A\alpha LB = \alpha^3 JALB, \qquad (1.40)$$

and the ratio is given by

$$\frac{F_\alpha}{F} = \alpha^3. \qquad (1.41)$$

1.7.1 Example: Compass Magnetometer

A magnetometer for a compass has to detect the Earth's magnetic field within the range of 0.25×10^{-4} T to 0.65×10^{4} T. Calculate the force produced by a value $B = 0.5 \times 10^{-4}$ T in a wire of length 2000 μm conducting an electrical current of intensity $I = 10$ mA.

Using equation (1.40),

$$F = 10 \times 10^{-3} \times 2000 \times 10^{-6} \times 0.5 \times 10^{-4} = 1 \times 10^{-9}\text{N}.$$

1.8　Dielectrophoretic Force

The dielectrophoretic force is the force that a non-uniform electric field exerts on a particle [6, p. 5]. The particle can be modelled as an electric dipole as shown in Figure 1.5.
　　The force acting on the dipole is

$$\vec{F} = Q\vec{E}(\vec{r} + \vec{d}) - Q\vec{E}(\vec{r}). \tag{1.42}$$

Linearizing the electric field by the first term of the Taylor expansion,

$$\vec{E}(\vec{r} + \vec{d}) = \vec{E} + \vec{d}\nabla\vec{E}(\vec{r}), \tag{1.43}$$

and substituting in equation (1.42),

$$\vec{F} = Q\vec{d}\nabla\vec{E}(\vec{r}). \tag{1.44}$$

In the limit we consider that when $r \to 0$ the dipole moment $\vec{p} = Q\vec{d}$ remains finite. According to a more detailed derivation in Chapter 6, the dipolar moment of a particle can be written as an effective dipole moment $p_{\text{eff}} \simeq Qd$, and then

$$\vec{F} = \overrightarrow{p_{\text{eff}}}\nabla\vec{E}. \tag{1.45}$$

The effective dipole moment is shown to be (see Chapter 7)

$$p_{\text{eff}} = 4\pi\epsilon_1 R^3 K\vec{E}, \tag{1.46}$$

where R is the particle radius, ϵ_1 is the medium permittivity and K is the Clausius–Mossotti factor

$$K = \frac{\epsilon_2 - \epsilon_1}{\epsilon_2 + 2\epsilon_1}, \tag{1.47}$$

in which ϵ_2 is the particle permittivity.

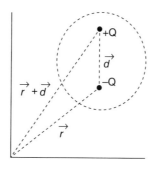

Figure 1.5　Dielectrophoretic force

The force on the particle is

$$\vec{F} = 4\pi\epsilon_1 R^3 K\vec{E}\nabla\vec{E}. \tag{1.48}$$

As the electric field is irrotational,

$$\vec{E}\nabla\vec{E} = \frac{1}{2}\nabla(\vec{E}\vec{E}) = \frac{1}{2}\nabla|\mathbf{E}|^2. \tag{1.49}$$

The dielectrophoretic force is

$$\vec{F} = 2\pi\epsilon_1 R^3 K\nabla|\mathbf{E}|^2. \tag{1.50}$$

Equation (1.50) indicates that the force has the same direction as the gradient of the square of the modulus of the electric field and depends on the third power of the particle radius. If we suppose, as an example, that the electric field is created by a charge $Q*$ located at the origin of our system coordinates, the electric field has spherical symmetry and is $\mathbf{E} = Q^*/r^2$.

The dielectrophoretic force on a sphere of radius R is then given by

$$\vec{F} = \frac{8\pi\epsilon_1 R^3 K Q^{*2}}{r^5}. \tag{1.51}$$

If the scaling scenario considers that the radius of the particle and the distance from the particle to the centre are equally scaled, then the scaling factor is

$$\frac{F_\alpha}{F} = \alpha^{-2}. \tag{1.52}$$

If the distance is not scaled, then

$$\frac{F_\alpha}{F} = \alpha^3. \tag{1.53}$$

1.8.1 Example: Nanoparticle in a Spherical Symmetry Electric Field

Consider a polystyrene nanoparticle of radius 300 nm ($\epsilon_r = 2.5$) and assume that $\nabla|\mathbf{E}|^2 = 3.3 \times 10^{23}$ V²/m³. Calculate the force at a distance 10 times the diameter of the sphere.

Taking into account that the Clausius–Mossotti factor is 0.333, the force is $F = 6.59 \times 10^{-9}$ N.

1.9 Summary

The different forces involved in microelectromechanical devices scale differently when the dimensions are scaled down. Looking at the α^n scaling law, the larger the value of n the more significantly the forces are reduced when the dimensions are reduced. As far as the forces examined in this chapter are concerned, the weight is the force that will become less and less important in the microworld. On the other hand, in the examples shown in this chapter we can also see that the forces present in common examples of today's MEMS devices vary quite widely in magnitude.

Table 1.1 shows a comparison of the scaling laws and a summary of the results of the examples worked in this chapter. It can be seen that the capillary and piezoelectric forces are quite significant (of the order of tenths of newtons), whereas magnetic and elastic forces, for the examples selected, are quite small.

Table 1.1 Summary of scaling laws and examples of the magnitude of forces

Force	Scaling law	Magnitude (N)	Example
Weight	α^3	2.85×10^{-5}	1.2.1
Elastic	α^2	8.1×10^{-8}	1.3.1
Electrostatic	α^0, α^2	8.96×10^{-7}	1.4.1
Capillary	α^1	4.8×10^{-1}	1.5.1
Piezoelectric	α^1, α^2	3.4×10^{-1}	1.6.1
Magnetic	α^1, α^3	1×10^{-9}	1.7.1
Dielctrophoresis	α^{-2}, α^3	6.59×10^{-9}	1.8.1

Problems

1.1 Calculate and plot the elastic restoring force of a cantilever having width $W = 10\,\mu m$ and thickness $h = 3\,\mu m$ and for lengths from $20\,\mu m$ to $2000\,\mu m$, when the deflection at the tip is 10% of the length. Take $E = 164 \times 10^9$ Pa.

1.2 For a silicon cantilever such as the one depicted in Figure 1.1, what is the most effective way to reduce the elastic constant k by a factor of 10 by changing only one of the dimensions? Similarly, what is the most effective way to increase k by a factor of 10?

1.3 We have an accelerometer based on an inertial mass from a cubic volume of silicon of $500\,\mu m$ side. Find the density of silicon and calculate the force that creates such mass when accelerated at 60 times gravity. If the inertial mass is supported by a flexure having an elastic constant of 100 N/m, find the mass displacement when the two forces reach equilibrium. If the edge of the cubic volume is at $5\,\mu m$ distance of a fixed electrode, find the capacitance value before the acceleration is applied to the mass and after. Assume that there is air in between the plates.

1.4 We have two plates of silver of area $250\,\mu m \times 250\,\mu m$ and $50\,\mu m$ thick. The upper plate is fixed and the bottom plate can move vertically. Calculate the minimum voltage that should be applied between the plates in order to start lifting the bottom plate.

1.5 We have two parallel electrodes at a distance of $4\,\mu m$ in air, with a voltage $V_{CC} = 10V$ applied between them. The permittivity ϵ_1 of air is equal to the permittivity of the vacuum ϵ_0. One of the electrodes is covered by a dielectric $2\,\mu m$ thick having a permittivity of $\epsilon_2 = \epsilon_0\epsilon_r$ with $\epsilon_r = 3.9$. Calculate and plot the electric field in the air and inside the dielectric.

1.6 A thin cylindrical capillary of 5 mm diameter is immersed in water. The surface tension is $\gamma = 72.8 \times 10^{-3}$ N/m, the liquid density is $\rho_m = 10^3$ kg/m^3 and the acceleration due to gravity is $g = 9.8$ m/s^2. Calculate the height of the water inside.

1.7 Compare the surface energy of a drop of liquid, assumed spherical in shape with radius r, with its volume.

1.8 We have a spherical drop of 2 mm radius. If the surface tension is $\gamma = 72.8 \times 10^{-3}$ N/m, calculate the surface energy change if the drop radius is stretched by Δr and find the change in the internal pressure in equilibrium.

1.9 A piezoelectric actuator has to produce a displacement in the bottom plate of a reservoir to eject droplets of ink to produce 600 dots per inch. Assume that the ink dot thickness is 1 μm and that there is just one drop per dot. Calculate the diameter of the dot, the volume of the drop and the radius of the drop (assumed to be equal to the radius of the ejecting nozzle). Calculate the vertical expansion required for a piezoelectric actuator acting on a cylindrical ink reservoir of 2 mm diameter. The thickness of the piezoelectric material is 10 μm.

1.10 We have a flexure made of gold and an electrical current of 10 mA circulates through it. If we immerse the flexure in a magnetic field normal to the plane of the flexure, calculate the force and indicate the direction of the movement. Take $L = 2000\,\mu m$, $B = 0.25 \times 10^{-4}$ T and the elastic constant of the flexure $k_x = 0.01$ N/m.

1.11 We have a 300 nm diameter polystyrene nanoparticle immersed in air, and an electric field created by a sphere carrying a total charge of Q. The distance between the particle and the sphere is 10 times the polystyrene sphere diameter. Calculate the force and the direction. Repeat the calculation if the medium is changed to ethylene glycol. Ethylene glycol has a relative permittivity of 37 and polystyrene 2.5.

2

Elasticity

In this chapter stress, strain and the relationship between them are described and formulated for isotropic and anisotropic materials. Miller indices are defined to identify crystallographic planes and to calculate angles between them. Stiffness and compliance matrices are introduced and Young's modulus and Poisson's ratio defined and calculated. Orthogonal transformation is used to find the values of the components of the stiffness and compliance matrices and of the values of the Young's modulus and Poisson's ratio when the coordinate system is rotated.

2.1 Stress

A body can be subject to external point forces, to applied force distributions and to support reactions. Stress is the combined effect of all these forces in a differential of area at a given point arbitrarily located in the body. Mathematically it is defined as the limit of the ratio of the differential resulting force $\vec{dF}$ divided by the differential of area $\vec{dA}$ [7, p. 201]:

$$T = \lim_{dA \to 0} \frac{\vec{dF}}{\vec{dA}}. \tag{2.1}$$

Both quantities are vectors, $\vec{dA}$ having the direction normal to the area $\vec{n}$, as shown in Figure 2.1.

Stress is the relationship existing between the components of the vector $\vec{dF}$ and the components of the vector $\vec{dA}$.

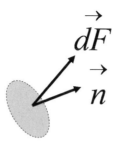

Figure 2.1 Definition of stress

Understanding MEMS: Principles and Applications, First Edition. Luis Castañer.
© 2016 John Wiley & Sons, Ltd. Published 2016 by John Wiley & Sons, Ltd.
Companion Website: www.wiley.com/go/castaner/understandingmems

In a one-dimensional geometry, such as shown in Figure 2.2, stress is defined as the ratio of the force applied to a surface divided by the area A. There are two possible uniaxial stresses depicted in the left and centre of Figure 2.2 where the force is tensile, tending to extend the length of the body (left), and compressive, where the force tends to reduce the length of the body (centre). The third case on the right of the figure is called shear stress, and the force in this case is tangential to the surface. In all three cases the stress is defined as

$$T = \frac{F}{A}. \tag{2.2}$$

More generally, in three dimensions and as the two vectors, force and differential of area, have three components each, the stress has nine components. The linear transformation of the components of one vector into another vector invariant to the rotation of the coordinate system is called a tensor of second rank (which is neither a scalar nor a vector). In matrix form,

$$\begin{pmatrix} dF_x \\ dF_y \\ dF_z \end{pmatrix} = \begin{pmatrix} dF_1 \\ dF_2 \\ dF_3 \end{pmatrix} = \begin{pmatrix} T_{11} & T_{12} & T_{13} \\ T_{21} & T_{22} & T_{23} \\ T_{31} & T_{32} & T_{33} \end{pmatrix} \begin{pmatrix} dA_1 \\ dA_2 \\ dA_2 \end{pmatrix}. \tag{2.3}$$

We will use numerical subindices instead of a cartesian nomenclature for the axes, and the correspondence is shown in Table 2.1.

In equation (2.3), T_{ij} designates the stress component of a force acting in the direction j on a surface having the normal in the direction of i. The uniaxial stresses are then those in the diagonal of the matrix and the other components are the shear stresses. For example, T_{21} is the shear stress originated by a force directed along axis 1 on a surface whose normal is directed along axis 2.

Considering a differential of volume as shown in Figure 2.3, there are 18 components of the stress in the six faces: six axial stresses and 12 shear stresses. For simplicity we show in Figure 2.3 the shear forces that create moments about the y-axis (or axis 2 in our nomenclature). The two stresses T'_{21} and T_{21} create moments about the centre in the clockwise direction, whereas the two stresses T'_{12} and T_{12} create moments in the counterclockwise direction.

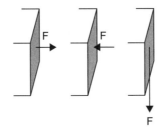

Figure 2.2   Axial and shear stresses

Table 2.1   Axis correspondence

Cartesian	Numerical
x	1
y	2
z	3

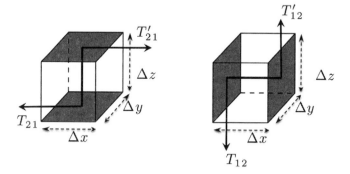

Figure 2.3   Shear stress components creating moments about the y-axis (or axis 2)

If the differential of volume has to be in equilibrium, then the total moment around the centre of the cube has to be zero:

$$T_{12}\Delta y\Delta z\frac{\Delta x}{2} + T'_{12}\Delta y\Delta z\frac{\Delta x}{2} - T'_{21}\Delta x\Delta y\frac{\Delta y}{2} - T_{21}\Delta x\Delta y\frac{\Delta y}{2} = 0. \tag{2.4}$$

If we assume that the dimensions are small, we can linearize

$$T'_{12} \simeq T_{12} + \frac{\partial T_{12}}{\partial x}\Delta x \quad \text{and} \quad T'_{21} \simeq T_{21} + \frac{\partial T_{21}}{\partial y}\Delta y, \tag{2.5}$$

and equation (2.4) becomes

$$2T_{12} - 2T_{21} + \frac{\partial T_{12}}{\partial x}\Delta x - \frac{\partial T_{21}}{\partial y}\Delta y = 0. \tag{2.6}$$

For $\Delta x \to 0$ and $\Delta y \to 0$,

$$T_{12} = T_{21}. \tag{2.7}$$

From similar calculations for the other directions of the moments it follows that $T_{13} = T_{31}$ and $T_{23} = T_{32}$.

The stress tensor components can then be defined by just six different magnitudes instead of nine: the axial components T_{11}, T_{22}, T_{33}, and the shear components T_{12}, T_{13} and T_{23}. Most of the time it is convenient to use a reduced tensor notation as shown in Table 2.2, where the correspondence to cartesian notation is also shown.

Table 2.2   Reduced tensor notation of stress components

Extended to reduced notation	Stress	Cartesian reduced notation
$11 \rightarrow 1$	$T_{11} \rightarrow T_1$	σ_x
$22 \rightarrow 2$	$T_{22} \rightarrow T_2$	σ_y
$33 \rightarrow 3$	$T_{33} \rightarrow T_3$	σ_z
$23 \rightarrow 4$	$T_{23} \rightarrow T_4$	τ_{yz}
$13 \rightarrow 5$	$T_{13} \rightarrow T_5$	τ_{xz}
$12 \rightarrow 6$	$T_{12} \rightarrow T_6$	τ_{xy}

2.2   Strain

The deformation of a body entails uniaxial deformation and pure shear deformation (not involving rotation). Let us first consider a simplified geometry, as depicted in Figure 2.4.

The strain S is defined as the relative deformation of the length of the body. This is shown in a two-dimensional geometry in Figure 2.4. The strain in the longitudinal direction is

$$S = \frac{u}{L}. \tag{2.8}$$

Due to the mass conservation law, if a body is stretched in one of the dimensions it should experience a contraction in the others. This is also shown in Figure 2.4.

Poisson's ratio, v, is defined as minus the ratio of the transversal and longitudinal strains,

$$v = -\frac{S_t}{S_l} = \frac{v/L}{u/L} = \frac{v}{u}. \tag{2.9}$$

More generally, the deformation can be defined by a displacement vector characterizing the change in the position of a point. This displacement vector is a function of the initial position. So we can define a second-rank tensor relating linearly the components of the displacement

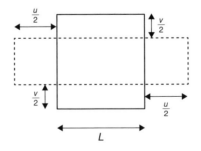

Figure 2.4   Longitudinal and transversal strains

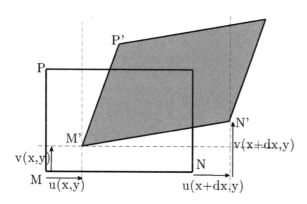

Figure 2.5   Definition of axial and shear strain

vector to the initial position vector. This tensor is called the strain. The components of strain are dimensionless.

The definition of the uniaxial and pure shear strain components is straightforward. Figure 2.5 shows an undeformed infinitesimal area and the same area after deformation (shaded). The horizontal and vertical components of the deformation vector (accounting for the displacement of corner points of the differential area) are $u(x, y)$ and $v(x, y)$.

Uniaxial strains along the x-axis and along the y-axis, S_{11} and S_{22}, are defined as

$$S_{11} = \frac{M'N' - MN}{MN} = \frac{\partial u}{\partial x}, \quad S_{22} = \frac{M'P' - MP}{MP} = \frac{\partial v}{\partial y}. \tag{2.10}$$

Also the engineering shear strain, S_{12}, is defined as the difference of interior angles of the body before and after deformation:

$$S_{12} = \frac{\pi}{2} - \left[\frac{\pi}{2} - \theta_1 - \theta_2\right], \quad S_{12} = \frac{\partial v}{\partial x} + \frac{\partial u}{\partial y} \tag{2.11}$$

The elastic shear strain used in many books is given by $S_{12}/2$. The definition of the strain components leads to symmetry,

$$\begin{pmatrix} du \\ dv \\ dw \end{pmatrix} = \begin{pmatrix} S_{11} & S_{12} & S_{13} \\ S_{21} & S_{22} & S_{23} \\ S_{31} & S_{32} & S_{33} \end{pmatrix} \begin{pmatrix} dx \\ dy \\ dz \end{pmatrix}, \tag{2.12}$$

with

$$S_{12} = S_{21}, \quad S_{13} = S_{31}, \quad S_{23} = S_{32}, \tag{2.13}$$

in such a way that the strain tensor has only six different components, $S_{11}, S_{22}, S_{33}, S_{23}, S_{13}, S_{12}$, or in reduced notation $S_1, S_2, S_3, S_4, S_5, S_6$, or again in reduced cartesian notation $\epsilon_x, \epsilon_y, \epsilon_z, \gamma_{yz}, \gamma_{xz}, \gamma_{xy}$. Table 2.3 shows the correspondence of the extended, short and cartesian notations for strain.

Table 2.3   Reduced tensor notation of strain components

Extended to reduced notation	Strain	Cartesian reduced notation
$11 \rightarrow 1$	$S_{11} \rightarrow S_1$	ϵ_x
$22 \rightarrow 2$	$S_{22} \rightarrow S_2$	ϵ_y
$33 \rightarrow 3$	$S_{33} \rightarrow S_3$	ϵ_z
$23 \rightarrow 4$	$S_{23} \rightarrow S_4$	γ_{yz}
$13 \rightarrow 5$	$S_{13} \rightarrow S_5$	γ_{xz}
$12 \rightarrow 6$	$S_{12} \rightarrow S_6$	γ_{xy}

2.3   Stress–strain Relationship

An elastic regime is defined when the deformations of the body are small and the stress and strain can be assumed to be proportional to each other,

$$E = \frac{T}{S}, \tag{2.14}$$

where E is Young's modulus or elasticity modulus and has units of pascals.

For isotropic materials, where the elastic properties are independent of the direction of the stress, in three dimensions the strain can be calculated as the linear superposition of the strains created by uniaxial stresses:

- $T_1 \neq 0$, $T_2 = 0$ and $T_3 = 0$. The strain in the longitudinal direction collinear with the stress T_1 is

$$S_1 = \frac{T_1}{E}, \tag{2.15}$$

and

$$S_2 = -\nu S_1, \quad S_3 = -\nu S_1. \tag{2.16}$$

Due to isotropy, no difference is expected between direction 2 and 3.
- $T_2 \neq 0$, $T_1 = 0$ and $T_3 = 0$. The strain in the longitudinal direction collinear with the stress T_2 is

$$S_2 = \frac{T_2}{E}, \tag{2.17}$$

and

$$S_1 = -\nu S_2, \quad S_3 = -\nu S_2. \tag{2.18}$$

- $T_3 \neq 0$, $T_1 = 0$ and $T_2 = 0$. The strain in the longitudinal direction collinear with the stress T_3 is

$$S_3 = \frac{T_3}{E}, \tag{2.19}$$

and

$$S_1 = -vS_3, \quad S_2 = -vS_3. \tag{2.20}$$

Applying linear superposition,

$$S_1 = \frac{1}{E}(T_1 - v(T_2 + T_3)), \tag{2.21}$$

$$S_2 = \frac{1}{E}(T_2 - v(T_1 + T_3)), \tag{2.22}$$

$$S_3 = \frac{1}{E}(T_3 - v(T_1 + T_2)). \tag{2.23}$$

The stress and strain shear components are related by a shear modulus G:

$$S_4 = \frac{1}{G}T_4, \tag{2.24}$$

$$S_5 = \frac{1}{G}T_5, \tag{2.25}$$

$$S_6 = \frac{1}{G}T_6, \tag{2.26}$$

with

$$G = \frac{E}{1+v}. \tag{2.27}$$

2.3.1 Example: Plane Stress

Plane stress is defined when one of the dimensions of the body is much smaller than the other two and the stress and strain out of the plane can be neglected. Suppose that $T_3 = S_3 = 0$. Write the stress in in the other two directions as a function of the in-plane stresses.

The equations are

$$S_1 = \frac{1}{E}(T_1 - vT_2), \tag{2.28}$$

$$S_2 = \frac{1}{E}(T_2 - vT_1). \tag{2.29}$$

$$\tag{2.30}$$

From these equations the stress components can be written as

$$T_1 = \frac{E}{1 - v^2}(S_1 + vS_2), \tag{2.31}$$

$$T_2 = \frac{E}{1 - v^2}(S_2 - vS_1). \tag{2.32}$$

$$\tag{2.33}$$

A further simplification can be done when the two stresses are equal, $T_1 = T_2 = T$ and $S_1 = S_2 = S$, leading to

$$T = \frac{E}{1 - v^2} S(1 + v) = \frac{E}{1 - v} S = DS, \tag{2.34}$$

where

$$D = \frac{E}{1 - v} \tag{2.35}$$

is called the biaxial modulus.

2.4 Strain–stress Relationship in Anisotropic Materials

In the case of an anisotropic material, as the stress and the strain tensors are of second rank (3×3 components), they are related by means of a fourth-rank tensor (9×9 components). However, as we have seen, both strain and stress tensors are symmetric and they only have six independent components each, so the notation can be simplified to

$$[S] = [s][T] \tag{2.36}$$

where $[s]$ is the compliance matrix. In full, we have

$$\begin{pmatrix} S_1 \\ S_2 \\ S_3 \\ S_4 \\ S_5 \\ S_6 \end{pmatrix} = \begin{pmatrix} s_{11} & s_{12} & s_{13} & s_{14} & s_{15} & s_{16} \\ s_{21} & s_{22} & s_{23} & s_{24} & s_{25} & s_{26} \\ s_{31} & s_{32} & s_{33} & s_{34} & s_{35} & s_{36} \\ s_{41} & s_{42} & s_{43} & s_{44} & s_{45} & s_{46} \\ s_{51} & s_{52} & s_{53} & s_{54} & s_{55} & s_{56} \\ s_{61} & s_{62} & s_{63} & s_{64} & s_{65} & s_{66} \end{pmatrix} \begin{pmatrix} T_1 \\ T_2 \\ T_3 \\ T_4 \\ T_5 \\ T_6 \end{pmatrix}. \tag{2.37}$$

The inverse of the compliance matrix is the stiffness matrix C with coefficients C_{ij},

$$[T] = [C][S] \tag{2.38}$$

or

$$\begin{pmatrix} T_1 \\ T_2 \\ T_3 \\ T_4 \\ T_5 \\ T_6 \end{pmatrix} = \begin{pmatrix} C_{11} & C_{12} & C_{13} & C_{14} & C_{15} & C_{16} \\ C_{21} & C_{22} & C_{23} & C_{24} & C_{25} & C_{26} \\ C_{31} & C_{32} & C_{33} & C_{34} & C_{35} & C_{36} \\ C_{41} & C_{42} & C_{43} & C_{44} & C_{45} & C_{46} \\ C_{51} & C_{52} & C_{53} & C_{54} & C_{55} & C_{56} \\ C_{61} & C_{62} & C_{63} & C_{64} & C_{65} & C_{66} \end{pmatrix} \begin{pmatrix} S_1 \\ S_2 \\ S_3 \\ S_4 \\ S_5 \\ S_6 \end{pmatrix}. \tag{2.39}$$

The reduced notation shown here can be expanded back to the full tensor notation just by using the same rules as in Table 2.2; for example $T_1 \rightarrow T_{11}$, $C_{11} \rightarrow C_{1111}$, $T_5 \rightarrow T_{13}$ and $C_{43} \rightarrow C_{2333}$.

Due to crystal symmetry there are fewer than 36 stiffness matrix coefficients; for example, even in the most asymmetrical triclinic crystal there are only 21 different coefficients.

2.5 Miller Indices

Crystalline materials such as silicon are often used in MEMS technology. The crystallographic planes and orientations are described by the Miller indices [8]. The procedure for calculating the Miller indices of a crystallographic plane is illustrated in Figure 2.6. The first step is to calculate the intersections of the plane with the crystallographic axis: a, b and c. The second step is to calculate the inverses of these three magnitudes: $1/a$, $1/b$ and $1/c$. The third and final step is to calculate the set of integer numbers in the same ratio with each other.

Figure 2.7 shows three orthogonal planes, each intersecting only one axis. The darker plane intersects the x-axis and hence the intersections are (a, ∞, ∞) and hence the Miller indices are (100). The lateral plane shaded lightest intersects the y-axis at a distance a from the origin and the two other axes at ∞. We can then write that the intersection set is (∞, a, ∞) and the set of reciprocal values $(0, 1/a, 0)$, and the closest set of integer numbers in the same ratio is (010). Finally, the top plane intersects only the z-axis and the Miller indices of this plane are (001).

The most common convention for the nomenclature is that a plane is identified by the Miller indices (hkl) in parentheses, whereas the direction normal to a given plane is identified by the same Miller indices but in brackets $[hkl]$. Moreover, due to the symmetries that many crystals have, all planes symmetrically equivalent to a given plane perpendicular to direction $[hkl]$ are described by $\{hkl\}$. Also by convention the x-axis is the direction [100], the y-axis the direction [010] and the z-axis the direction [001]. When the intersection of a plane with one of the axes lies in the negative direction, the corresponding Miller index

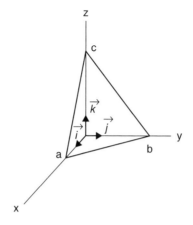

Figure 2.6   Miller indices definition

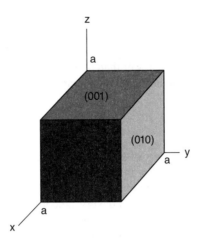

Figure 2.7   Milller indices of three orthogonal planes

is written with a bar on top; for example, $[h\bar{k}l]$ for a negative intersection with the y-axis. For example, for a cubic crystal which looks similar in all directions a set of equivalent planes is $(100), (\bar{1}00), (010), (0\bar{1}0), (001), (00\bar{1})$. The Miller index notation is summarized in Table 2.4.

Table 2.4   Miller index notation

Notation	Meaning
(hkl)	Plane in a crystal
$\{hkl\}$	Equivalent planes
$[hkl]$	Direction in a crystal
$\langle hkl \rangle$	Equivalent directions

2.5.1   Example: Miller Indices of Typical Planes

Find the Miller indices of the planes shown in Figure 2.8.

The plane shown in the left in Figure 2.8 intersects the x- and y-axes at a distance a from the origin. Hence the intersections are (a, a, ∞). Calculating the reciprocal values gives $(1/a, 1/a, 0)$, giving Miller indices (110). Similarly, the plane shown on the right of Figure 2.8 intersects the three axes at the same distance a, so the intersections are (a, a, a) and the reciprocal values $(1/a, 1/a, 1/a)$, giving the Miller indices (111).

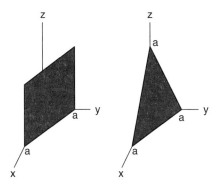

Figure 2.8   Example 2.5.1 of Miller indices

2.6   Angles of Crystallographic Planes

The Miller indices identify the planes and directions in a crystal. Crystal orientations are very important in MEMS technology, as key crystal properties are anisotropic in nature, such as Young's modulus. Miller indices serve to calculate the angle between two planes. If one of the planes has a normal $\vec{n}$ it can be written

$$\vec{n} = a\vec{i_1} + b\vec{i_2} + c\vec{i_3}, \tag{2.40}$$

where $(\vec{i_1}, \vec{i_2}, \vec{i_3})$ are the base unit vectors. The normal to the second plane $\vec{n'}$ can be written as

$$\vec{n'} = a'\vec{i_1} + b'\vec{i_2} + c'\vec{i_3} \tag{2.41}$$

in the same coordinate axes. The scalar product of two vectors is

$$\vec{n} \cdot \vec{n'} = |\vec{n}||\vec{n'}| \cos \theta_{nn'} = aa' + bb' + cc', \tag{2.42}$$

where $|\vec{n}|$ and $|\vec{n'}|$ are the moduli of the two normals. The cosine of the angle between the two normals is the same as the angle between the two planes and is given by

$$\cos \theta_{nn'} = \frac{aa' + bb' + cc'}{|\vec{n}||\vec{n'}|} = \frac{aa' + bb' + cc'}{\sqrt{a^2 + b^2 + c^2}\sqrt{a'^2 + b'^2 + c'^2}}. \tag{2.43}$$

2.6.1   Example

Calculate the angle between the planes identified by Miller indices (0,0,1) and (1,1,1).

We identify the planes by the normals,

$$\vec{n} = a\vec{i_1} + b\vec{i_2} + c\vec{i_3} = 1 \times \vec{i_1} + 0 \times \vec{i_2} + 0 \times \vec{i_3} = \vec{i_1},$$

$$\vec{n'} = a'\vec{i_1} + b'\vec{i_2} + c'\vec{i_3} = 1 \times \vec{i_1} + 1 \times \vec{i_2} + 1 \times \vec{i_3} = \vec{i_1} + \vec{i_2} + \vec{i_3}. \tag{2.44}$$

Applying equation (2.43),

$$\cos\theta_{nn'} = \frac{aa' + bb' + cc'}{\sqrt{a^2 + b^2 + c^2}\sqrt{a'^2 + b'^2 + c'^2}} = \frac{1 \times 1 + 0 \times 1 + 0 \times 1}{\sqrt{1^2 + 0^2 + 0^2}\sqrt{1^2 + 1^2 + 1^2}} = \frac{1}{\sqrt{3}}, \tag{2.45}$$

which corresponds to an angle of $\theta_{nn'} = \cos^{-1}(\frac{1}{\sqrt{3}}) = 54.74$ degrees.

2.7 Compliance and Stiffness Matrices for Single-Crystal Silicon

The stiffness matrix for silicon has only three independent non-zero components: $C_{11} = 166 \times 10^9$ Pa, $C_{12} = 64 \times 10^9$ Pa and $C_{44} = 80 \times 10^9$ Pa. Equation (2.39) simplifies to

$$\begin{pmatrix} T_1 \\ T_2 \\ T_3 \\ T_4 \\ T_5 \\ T_6 \end{pmatrix} = \begin{pmatrix} C_{11} & C_{12} & C_{12} & 0 & 0 & 0 \\ C_{12} & C_{11} & C_{12} & 0 & 0 & 0 \\ C_{12} & C_{12} & C_{11} & 0 & 0 & 0 \\ 0 & 0 & 0 & C_{44} & 0 & 0 \\ 0 & 0 & 0 & 0 & C_{44} & 0 \\ 0 & 0 & 0 & 0 & 0 & C_{44} \end{pmatrix} \begin{pmatrix} S_1 \\ S_2 \\ S_3 \\ S_4 \\ S_5 \\ S_6 \end{pmatrix} \tag{2.46}$$

or

$$\begin{pmatrix} T_1 \\ T_2 \\ T_3 \\ T_4 \\ T_5 \\ T_6 \end{pmatrix} = \begin{pmatrix} 166 & 64 & 64 & 0 & 0 & 0 \\ 64 & 166 & 64 & 0 & 0 & 0 \\ 64 & 64 & 166 & 0 & 0 & 0 \\ 0 & 0 & 0 & 80 & 0 & 0 \\ 0 & 0 & 0 & 0 & 80 & 0 \\ 0 & 0 & 0 & 0 & 0 & 80 \end{pmatrix} \begin{pmatrix} S_1 \\ S_2 \\ S_3 \\ S_4 \\ S_5 \\ S_6 \end{pmatrix}, \tag{2.47}$$

with the data values in gigapascals. Since the compliance matrix is the inverse of the stiffness matrix, we can also write the inverse of equation (2.46) as

$$\begin{pmatrix} S_1 \\ S_2 \\ S_3 \\ S_4 \\ S_5 \\ S_6 \end{pmatrix} = \begin{pmatrix} s_{11} & s_{12} & s_{12} & 0 & 0 & 0 \\ s_{12} & s_{11} & s_{12} & 0 & 0 & 0 \\ s_{12} & s_{12} & s_{11} & 0 & 0 & 0 \\ 0 & 0 & 0 & s_{44} & 0 & 0 \\ 0 & 0 & 0 & 0 & s_{44} & 0 \\ 0 & 0 & 0 & 0 & 0 & s_{44} \end{pmatrix} \begin{pmatrix} T_1 \\ T_2 \\ T_3 \\ T_4 \\ T_5 \\ T_6 \end{pmatrix}, \tag{2.48}$$

or numerically,

$$
\begin{pmatrix} S_1 \\ S_2 \\ S_3 \\ S_4 \\ S_5 \\ S_6 \end{pmatrix} = \begin{pmatrix} 0.0077 & -0.0021 & -0.0021 & 0 & 0 & 0 \\ -0.0021 & 0.0077 & -0.0021 & 0 & 0 & 0 \\ -0.0021 & -0.0021 & 0.0077 & 0 & 0 & 0 \\ 0 & 0 & 0 & 0.0125 & 0 & 0 \\ 0 & 0 & 0 & 0 & 0.0125 & 0 \\ 0 & 0 & 0 & 0 & 0 & 0.0125 \end{pmatrix} \begin{pmatrix} T_1 \\ T_2 \\ T_3 \\ T_4 \\ T_5 \\ T_6 \end{pmatrix}, \tag{2.49}
$$

with values in 10^{-9} Pa^{-1}.

The inversion of the compliance matrix can also be performed analytically as

$$
[s] = [C]^{-1} = \frac{1}{|C|}[\alpha^T], \tag{2.50}
$$

where $|C|$ is the determinant of the matrix C, and $[\alpha]^T$ is the transpose of the matrix of the co-factors of $[C]$. For example, the compliance matrix element s_{11} is given by

$$
s_{11} = \frac{1}{|C|} \begin{vmatrix} C_{11} & C_{12} & 0 & 0 & 0 \\ C_{12} & C_{11} & 0 & 0 & 0 \\ 0 & 0 & C_{44} & 0 & 0 \\ 0 & 0 & 0 & C_{44} & 0 \\ 0 & 0 & 0 & 0 & C_{44} \end{vmatrix} = \frac{1}{|C|} C_{44}^3 \left(C_{11}^2 - C_{12}^2 \right), \tag{2.51}
$$

and the determinant $|C|$ by

$$
|C| = C_{44}^3 \left(C_{11} \left(C_{11}^2 - 3C_{12}^2 \right) + 2C_{12}^3 \right), \tag{2.52}
$$

giving

$$
s_{11} = \frac{C_{11}^2 - C_{12}^2}{C_{11} \left(C_{11}^2 - 3C_{12}^2 \right) + 2C_{12}^3}. \tag{2.53}
$$

2.7.1   Example: Young's Modulus and Poisson Ratio for (100) Silicon

Calculate the values of Young's modulus and the Poisson ratio for the (100) direction.

Assuming that we only apply a uniaxial stress along the x-axis, only $T_1 \neq 0$. Then the strain–stress equation can be written as

$$
\begin{pmatrix} S_1 \\ S_2 \\ S_3 \\ S_4 \\ S_5 \\ S_6 \end{pmatrix} = \begin{pmatrix} s_{11} & s_{12} & s_{12} & 0 & 0 & 0 \\ s_{12} & s_{11} & s_{12} & 0 & 0 & 0 \\ s_{12} & s_{12} & s_{11} & 0 & 0 & 0 \\ 0 & 0 & 0 & s_{44} & 0 & 0 \\ 0 & 0 & 0 & 0 & s_{44} & 0 \\ 0 & 0 & 0 & 0 & 0 & s_{44} \end{pmatrix} \begin{pmatrix} T_1 \\ 0 \\ 0 \\ 0 \\ 0 \\ 0 \end{pmatrix}.
\tag{2.54}
$$

Young's modulus is defined as

$$
E_{100} = \frac{T_1}{S_1}.
\tag{2.55}
$$

From equation (2.54) it can be seen that

$$
E_{100} = \frac{1}{s_{11}}.
\tag{2.56}
$$

Taking into account equation (2.53),

$$
E_{100} = \frac{1}{s_{11}} = C_{11} - 2\frac{C_{12}^2}{C_{11} + C_{12}} = 130\,\text{GPa}.
\tag{2.57}
$$

This result is the same for the axial stresses along the other two axes as the three axes are equivalent. The Poisson ratio is defined as the ratio of the transversal to longitudinal strains. If we consider a geometry where the longitudinal strain is S_1 and the transversal strain is S_2, then from equation (2.54),

$$
v = -\frac{S_2}{S_1} = -\frac{s_{12}T_1}{s_{11}T_1} = -\frac{s_{12}}{s_{11}}.
\tag{2.58}
$$

The compliance matrix coefficient s_{12} is

$$
s_{12} = -\frac{1}{|C|} \begin{vmatrix} C_{12} & C_{12} & 0 & 0 & 0 \\ C_{12} & C_{11} & 0 & 0 & 0 \\ 0 & 0 & C_{44} & 0 & 0 \\ 0 & 0 & 0 & C_{44} & 0 \\ 0 & 0 & 0 & 0 & C_{44} \end{vmatrix} = -\frac{1}{|C|} C_{44}^3 \left(C_{11}C_{12} - C_{12}^2 \right),
\tag{2.59}
$$

and the Poisson ratio is

$$
v = -\frac{S_2}{S_1} = \frac{C_{11}C_{12} - C_{12}^2}{C_{11}^2 - C_{12}^2} = \frac{C_{12}}{C_{11} + C_{12}}
\tag{2.60}
$$

with value

$$v = -\frac{s_{12}}{s_{11}} = \frac{0.0021}{0.0077} = 0.272. \qquad (2.61)$$

It is also easy to show that the shear components of the stress and strain are related by the shear modulus G:

$$G = \frac{T_4}{S_4} = \frac{T_5}{S_5} = \frac{T_6}{S_6} = C_{44}. \qquad (2.62)$$

2.8 Orthogonal Transformation

In the design of microelectromechanical devices the coordinate system of forces does not generally coincide with the coordinate system where the elastic properties of the materials are known. There is a need to transform the stiffness and compliance matrices from the crystallographic coordinate system into the application coordinate system. This is done by means of an orthogonal transformation [9]. Given an orthogonal coordinate system base of unit vectors $(\vec{i}_1, \vec{i}_2, \vec{i}_3)$, a vector $\vec{A}$ can be written as

$$\vec{A} = A_1\vec{i}_1 + A_2\vec{i}_2 + A_3\vec{i}_3, \quad i'_j \cdot i_k = \delta_{ij}, \ A_i = \vec{A} \cdot \vec{i}_i \qquad (2.63)$$

where $\delta_{jk} = 1$ if $i = j$ and $\delta_{jk} = 0$ if $i \neq j$.

The same vector $\vec{A}$ can be written as a function of a different base of unit vectors belonging to a rotated coordinate system $(\vec{i}'_1, \vec{i}'_2, \vec{i}'_3)$,

$$\vec{A} = A'_1\vec{i}'_1 + A'_2\vec{i}'_2 + A'_3\vec{i}'_3, \quad i'_j \cdot i'_k = \delta_{jk}, \ A'_i = \vec{A} \cdot \vec{i}'_i. \qquad (2.64)$$

Multiplying equation (2.64) on the right by $\vec{i}'_1$,

$$\vec{A}\vec{i}'_1 = A'_1\vec{i}'_1\vec{i}'_1 + A'_2\vec{i}'_2\vec{i}'_1 + A'_3\vec{i}'_3\vec{i}'_1 = A'_1 \qquad (2.65)$$

Multiplying equation (2.63) by $\vec{i}'_1$

$$\vec{A}\vec{i}'_1 = A_1\vec{i}_1\vec{i}'_1 + A_2\vec{i}_2\vec{i}'_1 + A_3\vec{i}_3\vec{i}'_1. \qquad (2.66)$$

As the vector $\vec{A}$ is the same in the two coordinate systems, equations (2.65) and (2.66) must be equal,

$$A'_1 = A_1\vec{i}_1\vec{i}'_1 + A_2\vec{i}_2\vec{i}'_1 + A_3\vec{i}_3\vec{i}'_1, \qquad (2.67)$$

so in general

$$A'_j = A_1 \vec{i_1} \vec{i'_j} + A_2 \vec{i_2} \vec{i'_j} + A_3 \vec{i_3} \vec{i'_j}. \tag{2.68}$$

As can be seen, the scalar products of the unit vectors in the two coordinate systems are involved in equation (2.68). To simplify notation we define a_{jk} as

$$a_{jk} = \vec{i_j} \cdot \vec{i'_k} = |\vec{i_j}| \cdot |\vec{i'_k}| \cos\theta_{i_j i'_k} = \cos\theta_{i_j i'_k}; \tag{2.69}$$

a_{jk} are then the director cosines of the angles between the unit vectors $(\vec{i_1}, \vec{i_2}, \vec{i_3})$ and the unit vectors $(\vec{i'_1}, \vec{i'_2}, \vec{i'_3})$.

The components of the vector $\vec{A}$ in the rotated coordinate system are

$$A'_j = \vec{A} \vec{i'_j} = A_1 i_1 i'_j + A_2 i_2 i'_j + A_3 i_3 i'_j = \sum_{k=1}^{k=3} a_{jk} A_k. \tag{2.70}$$

In matrix notation equation (2.70) can be written as

$$\begin{pmatrix} A'_1 \\ A'_2 \\ A'_3 \end{pmatrix} = (R) \begin{pmatrix} A_1 \\ A_2 \\ A_3 \end{pmatrix}. \tag{2.71}$$

(R) is the matrix of director cosines and is called the rotation matrix. In many books the notation $a_{11} = l_1, a_{12} = m_1, a_{13} = n_1$, etc. is used, so

$$(R) = \begin{pmatrix} \cos\theta_{i'_1 i_1} & \cos\theta_{i'_1 i_2} & \cos\theta_{i'_1 i_3} \\ \cos\theta_{i'_2 i_1} & \cos\theta_{i'_2 i_2} & \cos\theta_{i'_2 i_3} \\ \cos\theta_{i'_3 i_1} & \cos\theta_{i'_3 i_2} & \cos\theta_{i'_3 i_3} \end{pmatrix} = \begin{pmatrix} a_{11} & a_{12} & a_{13} \\ a_{21} & a_{22} & a_{23} \\ a_{31} & a_{32} & a_{33} \end{pmatrix} = \begin{pmatrix} l_1 & m_1 & n_1 \\ l_2 & m_2 & n_2 \\ l_3 & m_3 & n_3 \end{pmatrix}. \tag{2.72}$$

The rotation matrix has the property that its inverse and its transpose are equal, $R^T = R^{-1}$, leading to the following equations,

$$l_1^2 + m_1^2 + n_1^2 = 1,$$
$$l_2^2 + m_2^2 + n_2^2 = 1,$$
$$l_3^2 + m_3^2 + n_3^2 = 1,$$
$$l_1 l_2 + m_1 m_2 + n_1 n_2 = 0,$$
$$l_1 l_3 + m_1 m_3 + n_1 n_3 = 0,$$
$$l_2 l_3 + m_2 m_3 + n_2 n_3 = 0. \tag{2.73}$$

2.9   Transformation of the Stress State

The stress tensor in a rotated coordinate system by means of an orthogonal transformation T'_{il} can be related to the stress tensor components in the unrotated system, T_{jk}, as follows [10, p. 17]:

$$T'_{il} = \sum_{j=1}^{j=3} \sum_{k=1}^{k=3} a_{ij} a_{lk} T_{lk}, \tag{2.74}$$

In matrix notation, equation (2.74) can be written as

$$(T') = (R)(T)(R)^T. \tag{2.75}$$

2.9.1   Example: Rotation of the Stress State

The stress state in a given element of a body is known and it is given by the stress tensor (T, in megapascals) as follows:

$$(T) = \begin{pmatrix} -10 & 5 & -1 \\ 5 & 2 & 3 \\ -1 & 3 & -4 \end{pmatrix}. \tag{2.76}$$

Calculate the stress tensor if the original axes are rotated by an angle $\alpha = 45°$ about the z-axis.

Let us assume that the z-axis corresponds to the unit vector $\vec{i}_3$. Then the angles between the original and rotated axes are

$$\begin{pmatrix} \theta_{i'_1 i_1} = \alpha & \theta_{i'_1 i_2} = \frac{\pi}{2} - \alpha & \theta_{i'_1 i_3} = \frac{\pi}{2} \\ \theta_{i'_2 i_1} = \frac{\pi}{2} + \alpha & \theta_{i'_2 i_2} = \alpha & \theta_{i'_2 i_3} = \frac{\pi}{2} \\ \theta_{i'_3 i_1} = \frac{\pi}{2} & \theta_{i'_3 i_2} = \frac{\pi}{2} & \theta_{i'_3 i_3} = 0 \end{pmatrix}. \tag{2.77}$$

So the rotation matrix is

$$(R) = \begin{pmatrix} \cos\alpha & \cos(\frac{\pi}{2} - \alpha) & 0 \\ \cos(\frac{\pi}{2} + \alpha) & \cos\alpha & 0 \\ 0 & 0 & 1 \end{pmatrix} = \begin{pmatrix} \cos\alpha & \sin\alpha & 0 \\ -\sin\alpha & \cos\alpha & 0 \\ 0 & 0 & 1 \end{pmatrix} \tag{2.78}$$

$$= \begin{pmatrix} \sqrt{0.5} & \sqrt{0.5} & 0 \\ -\sqrt{0.5} & \sqrt{0.5} & 0 \\ 0 & 0 & 1 \end{pmatrix}.$$

The stress tensor in the rotated axis is

$$(T') = (R)(T)(R)^T \tag{2.79}$$

$$= \begin{pmatrix} \cos\alpha & \sin\alpha & 0 \\ -\sin\alpha & \cos\alpha & 0 \\ 0 & 0 & 1 \end{pmatrix} \begin{pmatrix} -10 & 5 & -1 \\ 5 & 2 & 3 \\ -1 & 3 & -4 \end{pmatrix} \begin{pmatrix} \cos\alpha & -\sin\alpha & 0 \\ \sin\alpha & \cos\alpha & 0 \\ 0 & 0 & 1 \end{pmatrix}$$

$$= \begin{pmatrix} 1 & 6 & \sqrt{2} \\ 6 & -9 & 2\sqrt{2} \\ \sqrt{2} & 2\sqrt{2} & -4 \end{pmatrix}.$$

2.9.2 Example: Matrix Notation for the Rotation of the Stress State

For the same data as in the example above, calculate the value of the tensor element T_{11} using the index notation.

We have

$$T'_{11} = \sum_{j=1}^{j=3} \sum_{k=1}^{k=3} a_{ij} a_{lk} T_{lk} \tag{2.80}$$

$$\begin{aligned} = \; & a_{11}a_{11}T_{11} + a_{11}a_{12}T_{12} + a_{11}a_{13}T_{13} + a_{12}a_{11}T_{21} \\ & + a_{12}a_{12}T_{22} + a_{12}a_{13}T_{23} + a_{13}a_{11}T_{31} + a_{13}a_{12}T_{32} + a_{13}a_{13}T_{33}. \end{aligned} \tag{2.81}$$

Making use of equations (2.72) and (2.78) to find the values of the directional cosines a_{ij}, it follows that

$$\begin{aligned} T'_{11} = \; & 0.707 \times 0.707 \times (-10) + 0.707 \times 0.707 \times 5 + 0.707 \times 0 \times (-1) - \\ & -0.707 \times 0.707 \times 5 + 0.707 \times 0.707 \times 2 - 0.707 \times 0 \times 3 + 0 + 0 + 0 \\ = \; & 0.999, \end{aligned} \tag{2.82}$$

which is the same result as in the example above.

2.10 Orthogonal Transformation of the Stiffness Matrix

The stiffness matrix is a fourth-rank tensor which under an orthogonal transformation from one set of axes to another in a rotation becomes [11]

$$C'_{ijkl} = \sum_{a=1}^{a=3} \sum_{b=1}^{b=3} \sum_{c=1}^{c=3} \sum_{d=1}^{d=3} a_{ia} a_{jb} a_{kc} a_{ld} C_{abcd}. \tag{2.83}$$

There are also analytical equations relating the rotated components of the stiffness matrix with the values in the crystallographic coordinate system and to the director cosines [12]. The equations for C'_{11} and for C'_{12} are reproduced here:

$$C'_{11} = C_{11} + C_c \left(l_1^4 + m_1^4 + n_1^4 - 1 \right),$$
$$C'_{12} = C_{12} + C_c \left(l_1^2 l_2^2 + m_1^2 m_2^2 + n_1^2 n_2^2 \right),$$

where

$$C_c = C_{11} - C_{12} - 2C_{44}. \tag{2.84}$$

Similarly, in the same reference [12] analytical equations for the compliance matrix coefficients in a rotated coordinate system are described. The equations for s'_{11} and for s'_{12} are reproduced here:

$$s'_{11} = s_{11} + s_c \left(l_1^4 + m_1^4 + n_1^4 - 1 \right),$$
$$s'_{12} = s_{12} + s_c \left(l_1^2 l_2^2 + m_1^2 m_2^2 + n_1^2 n_2^2 \right),$$

where

$$s_c = s_{11} - s_{12} - \frac{1}{2}C_{44}. \tag{2.85}$$

2.10.1 Example: C_{11} Coefficient in Rotated Axes

Show that the C_{11} coefficient in rotated axes is given by

$$C'_{11} = C_{11} - \left(C_{11} - C_{12} - 2C_{44} \right)\left(l_1^4 + m_1^4 + n_1^4 - 1 \right). \tag{2.86}$$

The subindices of the stiffness coefficient C_{11} can be expanded according to Table 2.2 to give C'_{1111} and equation (2.83) can be written

$$
\begin{aligned}
C'_{1111} = \ & a_{11}a_{11}a_{11}a_{11}C_{1111} + a_{12}a_{12}a_{11}a_{11}C_{2211} + a_{13}a_{13}a_{11}a_{11}C_{3311} + a_{12}a_{13}a_{11}a_{11}C_{2311} \\
& + a_{13}a_{12}a_{11}a_{11}C_{3211} + a_{11}a_{13}a_{11}a_{11}C_{1311} + a_{13}a_{11}a_{11}a_{11}C_{3111} + a_{11}a_{12}a_{11}a_{11}C_{1211} \\
& + a_{12}a_{11}a_{11}a_{11}C_{2111} + a_{11}a_{11}a_{12}a_{12}C_{1122} + a_{12}a_{12}a_{12}a_{12}C_{2222} + a_{13}a_{13}a_{12}a_{12}C_{3322} \\
& + a_{11}a_{11}a_{13}a_{13}C_{1133} + a_{12}a_{12}a_{13}a_{13}C_{2233} + a_{13}a_{13}a_{13}a_{13}C_{3333} + a_{12}a_{13}a_{12}a_{13}C_{2323} \\
& + a_{13}a_{12}a_{12}a_{13}C_{3223} + a_{12}a_{13}a_{13}a_{12}C_{2332} + a_{13}a_{12}a_{13}a_{12}C_{3232} + a_{11}a_{13}a_{11}a_{13}C_{1313} \\
& + a_{13}a_{11}a_{11}a_{13}C_{3113} + a_{11}a_{13}a_{13}a_{11}C_{1331} + a_{13}a_{11}a_{13}a_{11}C_{3131} + a_{11}a_{12}a_{11}a_{12}C_{1212} \\
& + a_{12}a_{11}a_{11}a_{12}C_{2112} + a_{11}a_{12}a_{12}a_{11}C_{1221} + a_{12}a_{11}a_{12}a_{11}C_{2121}.
\end{aligned}
$$

Using the reduced notation for the unrotated coefficients shown in Table 2.2 (e.g. $C_{1111} \rightarrow C_{11}$ or $C_{3232} \rightarrow C_{44}$) and taking into account that many of the elements of the stiffness matrix are zero and that those that are not zero take only three different values, C_{11}, C_{12} or C_{44}, it follows that

$$C'_{1111} = C_{11} \left(l_1^4 + m_1^4 + n_1^4 \right) + \left(2C_{12} + 4C_{44} \right) \left(m_1^2 l_1^2 + n_1^2 l_1^2 + m_1^2 n_1^2 \right). \tag{2.87}$$

If we take into account that the direction cosines satisfy the equation

$$l_1^2 + m_1^2 + n_1^2 = 1, \tag{2.88}$$

then

$$\left(l_1^2 + m_1^2 + n_1^2 \right)^2 = l_1^4 + m_1^4 + n_1^4 + 2 \left(m_1^2 l_1^2 + n_1^2 l_1^2 + m_1^2 n_1^2 \right) = 1. \tag{2.89}$$

Then

$$2 \left(m_1^2 l_1^2 + n_1^2 l_1^2 + m_1^2 n_1^2 \right) = 1 - \left(l_1^4 + m_1^4 + n_1^4 \right). \tag{2.90}$$

Equation (2.87) becomes

$$C'_{1111} = (C_{11} - C_{12} - 2C_{44}) \left(l_1^4 + m_1^4 + n_1^4 \right) + C_{12} + 2C_{44}. \tag{2.91}$$

Adding and subtracting C_{11},

$$C'_{1111} = (C_{11} - C_{12} - 2C_{44}) \left(l_1^4 + m_1^4 + n_1^4 \right) + C_{12} + 2C_{44} + C_{11} - C_{11} \tag{2.92}$$

leading to the result,

$$C'_{11} = C_{11} - (C_{11} - C_{12} - 2C_{44}) \left(l_1^4 + m_1^4 + n_1^4 - 1 \right). \tag{2.93}$$

2.10.2   Example: Young's Modulus and Poisson Ratio in the (111) Direction

Choose a set of three coordinate axes orthogonal to each other and including the axis (111). Calculate the rotation matrix between the crystallographic axes and the new axes. Calculate Young's modulus and the Poisson ratio in the (111) direction

---

The set of directions $\vec{i'_1} = (111)$, $\vec{i'_2} = (0\bar{1}1)$ and $\vec{i'_3} = (2\bar{1}\bar{1})$ are mutually orthogonal. For example, it can be seen that (111) and $(0\bar{1}1)$ are orthogonal because

$$\cos \theta = \frac{aa' + bb' + cc'}{\sqrt{a^2 + b^2 + c^2}\sqrt{a'^2 + b'^2 + c'^2}} = \frac{0 - 1 + 1}{\sqrt{3}\sqrt{2}} = 0, \tag{2.94}$$

and then $\theta = \pi/2$.

The rotation matrix between the set of axes (XYZ) $(\vec{i_1}\vec{i_2}\vec{i_3})$ and $(X'Y'Z')$, $(\vec{i_1'}\vec{i_2'}\vec{i_3'})$ is given by

$$(R) = \begin{pmatrix} \cos\theta_{i_1'i_1} & \cos\theta_{i_1'i_2} & \cos\theta_{i_1'i_3} \\ \cos\theta_{i_2'i_1} & \cos\theta_{i_2'i_2} & \cos\theta_{i_2'i_3} \\ \cos\theta_{i_3'i_1} & \cos\theta_{i_3'i_2} & \cos\theta_{i_3'i_3} \end{pmatrix} = \begin{pmatrix} l_1 & m_1 & n_1 \\ l_2 & m_2 & n_2 \\ l_3 & m_3 & n_3 \end{pmatrix} = \begin{pmatrix} \frac{1}{\sqrt{3}} & \frac{1}{\sqrt{3}} & \frac{1}{\sqrt{3}} \\ \frac{1}{\sqrt{2}} & \frac{-1}{\sqrt{2}} & \frac{-1}{\sqrt{2}} \\ \frac{-1}{\sqrt{6}} & \frac{-1}{\sqrt{6}} & \frac{2}{\sqrt{2}} \end{pmatrix}. \tag{2.95}$$

Young's modulus E_{111} in the direction (111) of the rotated axis is given by the same equation (2.56) but in the rotated axes. Taking into account that the direction [111] has been assigned to the unit vector $\vec{i_1'}$ that corresponds to the X' axis,

$$E_{111} = \frac{T_1'}{S_1'} = \frac{1}{s_{11}'}. \tag{2.96}$$

We have

$$s_{11}' = s_{11} + \left(s_{11} - s_{12} - \frac{1}{2}s_{44}\right)\left(l_1^4 + m_1^4 + n_1^4 - 1\right) = 0.00534 \tag{2.97}$$

and

$$E_{111} = \frac{1}{s_{11}'} = \frac{1}{0.00534} = 187.2\,\text{GPa}. \tag{2.98}$$

The Poisson ratio is defined similarly to the equation (2.60) in the rotated axes, as the longitudinal direction is assigned to the X' axis and the transversal direction to the Y' axis,

$$v_{111} = -\frac{S_2'}{S_1'} = -\frac{s_{12}'}{s_{11}'}. \tag{2.99}$$

We have

$$s_{12}' = s_{12} + \left(s_{11} - s_{12} - \frac{1}{2}s_{44}\right)\left(l_1^2 l_2^2 + m_1^2 m_2^2 + n_1^2 n_2^2\right) = -0.00075 \tag{2.100}$$

and

$$v_{111} = -\frac{-0.00075}{0.00534} = 0.12. \tag{2.101}$$

2.11   Elastic Properties of Selected MEMS Materials

Several materials are used as structural materials in MEMS devices, besides single-crystal silicon. They are summarized in Table 2.5.

Table 2.5   Elastic properties of selected materials

	Young's modulus (GPa)	Poisson's ratio	Shear modulus (GPa)	Ref.
Silicon nitride	166–297	0.23–0.28	65.3–127	[13]
Silicon dioxide	66.3–74.8	0.15–0.19	27.9–32.3	[13]
Polysilicon	169	0.22	69	[14]

Problems

2.1   Find suitable axes for a silicon wafer with orientation (100), and demonstrate that the three unit vectors are orthogonal.

2.2   Using the set of unit vectors found in Problem 2.1, calculate the rotation matrix from the crystallographic axes to the new ones.

2.3   Using the set of unit vectors found in Problem 2.1, calculate the values of Young's modulus and the Poisson ratio in the rotated axes.

2.4   Using the set of unit vectors found in Problem 2.1, calculate the values of the elements C'_{11} and C'_{12} of the stiffness matrix in the rotated axes.

2.5   Write Matlab code and find the values of all elements of the stiffness matrix in the rotated system for Problem 2.1.

2.6   Using the set of unit vectors found in Problem 2.1, calculate the values of the elements of the compliance matrix in the rotated axes.

2.7   Show that the element C'_{12} of the stiffness matrix in the rotated axes is given by [12]

$$C'_{12} = C_{12} + (C_{11} - C_{12} - 2C_{44}) \left(l_1^2 l_2^2 + m_1^2 m_2^2 + n_1^2 n_2^2 \right). \qquad (2.102)$$

2.8   Calculate the rotation matrix when the crystallographic axes are rotated around the z-axis by an angle β.

2.9   Write Matlab code or similar to calculate the values of the elements C'_{11} and C'_{12} of the rotated stiffness matrix when the rotation is around the z-axis and with angles of 15°, 30°, 45°, 60° and 90°.

2.10   Calculate the values of Young's modulus and the Poisson ratio in the plane when the rotation is around the z-axis and with angles of 15°, 30° and 45°.

3

Bending of Microstructures

The operation of a large variety of MEMS devices is based on the deformation of microstructures subject to forces. In this chapter the basics of the bending of microstructures such as cantilevers, bridges, flexures and membranes is covered with the aim of developing and analytically solving the main equations that allow us to calculate the deflection of the microstructures. The main steps are:

- a reminder of the static equilibrium conditions and the free body diagram for calculating bending moments and reaction and shear forces;
- definition of the neutral surface;
- definition and calculation of the moment of inertia;
- formulation of the beam differential equation relating the deflection to the bending moment;
- solution of the beam equation in examples with typical boundary conditions;
- calculation of the stiffness or elastic constant for microstructures and especially for flexures applying Castigliano's second theorem;
- formulation of the rectangular and circular plate differential equation and solution when pressure is applied.

In this chapter simplifications are made to enable analytical solutions to be generated.

3.1 Static Equilibrium

A body is in static equilibrium when the net applied force and the net moment are zero:

$$\sum \mathbf{F} = 0, \quad \sum \mathbf{M} = 0. \tag{3.1}$$

Application of static equilibrium conditions allows us to calculate reaction forces and moments as well as shear forces and moments in specific sections of a microstructure. Additionally, the solution of the differential equations for beams and membranes requires the application of boundary conditions depending on the type of support, according to the definitions shown in Figure 3.1. As can be seen in the enlarged view of the end or support of a generic beam,

Understanding MEMS: Principles and Applications, First Edition. Luis Castañer.
© 2016 John Wiley & Sons, Ltd. Published 2016 by John Wiley & Sons, Ltd.
Companion Website: www.wiley.com/go/castaner/understandingmems

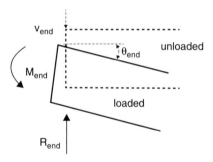

Figure 3.1   Enlarged view of the end of a generic beam

a reaction force R_{end}, a reaction moment M_{end}, an end angle θ_{end} and an end displacement or deflection v_{end} compared to the unloaded beam are defined.

The end or support of a beam can be of various kinds:

- Fixed or clamped: the beam at the support cannot move, either horizontally or vertically, and the slope at the support should be zero.
- Simple: same conditions as for fixed supports, with the exception that the slope at the support can take any finite value.
- Guided end: a fixed angle is required at the support.

Table 3.1 summarizes the boundary mathematical conditions that apply to the supports compared to a free end [10, p. 17]. For example, for a fixed or clamped support, the requirements are that the deflection and the slope are zero.

Table 3.1   Beam end restraints for different supports

Beam end	Restraints	
Free	$M_{end} = 0$	$R_{end} = 0$
Simple	$M_{end} = 0$	$v_{end} = 0$
Fixed or clamped	$v_{end} = 0$	$\theta_{end} = 0$
Guided	$R_{end} = 0$	$\theta_{end} = 0$

3.2   Free Body Diagram

In MEMS devices we are mostly interested in elastic structures that deform under external forces or distributed loads, and in particular in finding the value of the deformation at a given point. A differential beam equation has to be solved where the deflection $v(x)$ is related to the applied forces. If we consider, for example, a beam subject to an external force pushing down the beam, deformation will contract the top part of the beam and expand the bottom part. This creates a distribution of stress in the cross-section.

For a body in static equilibrium, any part of it must also be in equilibrium. One useful way to isolate a part of a body to study forces and moments is to cut that part of the body and

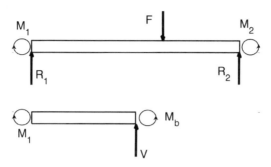

Figure 3.2   Free body diagram, point external force and reactions at the supports (top), and left section at x (bottom)

replace the effects of external forces and of the rest of the body by a shear force and a bending moment, as shown in Figure 3.2. This is known as a free body diagram.

Figure 3.2 shows an example of a beam supported at the two ends where a point force F is applied. In order to preserve equilibrium, there are reaction forces at the supports R_1 and R_2 and reaction moments M_1 and M_2. Writing force and moment equilibrium equations allows us to find relationships between the reaction forces and moments with the external forces. If the beam is in equilibrium, any section of must also be in equilibrium. In Figure 3.2 we consider the left section of the beam, from the left end to the position x, and consider it to be isolated from the rest. For the equilibrium to be preserved, and additional shear force V and bending moment M_b have to be postulated. The values of V and M_b can be calculated as a function of the external forces and reactions.

From the equilibrium equations applied to a section of the beam we can find a function of the moments and forces and of the bending moment at any section x.

3.3   Neutral Plane and Curvature

We will consider a deformed beam as shown in Figure 3.3 where an unbent (left) and bent (right) beam can be seen. When a beam is bent as shown, the top planes will be compressed and the bottom planes will be subject to expansion. Somewhere in between there should be a plane that is not deformed. This plane is defined as the neutral plane. In Figure 3.3 the planes

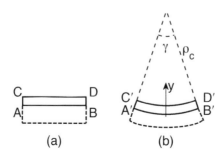

Figure 3.3   Bending of a beam: (a) unbent, (b) after bending

above the neutral plane are shown in solid black, whereas the planes below the neutral plane are shown dashed. We concentrate on the planes above the neutral plane. We will suppose that after bending the the cross-section shown can be approximated by circular sectors having a radius of curvature ρ_c.

Following a derivation similar to that in reference [7, p. 421], we consider the segment AB belonging to the neutral plane. Its length will be the same after bending, hence $AB = A'B'$, whereas the segment CD which is above of the neutral plane changes its length to $C'D' \neq CD$. It is straightforward to relate the change in length to the radius of curvature. Taking into account the definition of axial strain ϵ_x as the relative change in length of segment CD,

$$\epsilon_x = \frac{C'D' - CD}{CD} = \frac{C'D' - AB}{AB} = \frac{(\rho_c - y)\gamma - \rho_c\gamma}{\rho_c\gamma}, \tag{3.2}$$

we have

$$\epsilon_x = -\frac{y}{\rho_c}. \tag{3.3}$$

In this chapter the cartesian reduced notation described in Table 2.3 is used because numerical indices are required for numerical calculations; for example, $\epsilon_x = S_{11}$ in extended numerical notation or $\epsilon_x = S_1$ in numerical short notation.

Equation (3.2) provides the value of the strain at any given position y inside the beam. If we suppose that the medium is isotropic, the uniaxial stress σ_x is given by

$$\sigma_x = -E\frac{y}{\rho_c}. \tag{3.4}$$

E is the Young's modulus corresponding to this axial direction. From equation (3.4) the stress is negative for $y > 0$ (above the neutral plane) with sign opposite to the x-axis. In contrast, the stress is positive for $y < 0$ (below the neutral plane), consistent with tensile stress.

3.4   Pure Bending

We define the conditions of pure bending as the situation where there is no net axial force in an arbitrarily chosen cross-section. According to Figure 3.4,

$$\sum F_x = 0 = \int_A \sigma_x dA. \tag{3.5}$$

Using equation (3.4),

$$\int_A \frac{E}{\rho} y dA = 0, \tag{3.6}$$

and we have

$$\int_A y dA = 0. \tag{3.7}$$

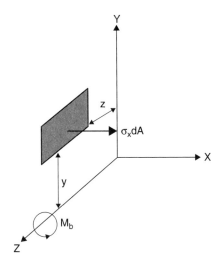

Figure 3.4   Forces and moments in a cross-section

The integral in equation (3.6) is known as the first moment of cross-sectional area about the neutral surface, and when it is equal to zero it allows the calculation of the position of the neutral surface for a given cross-section.

3.4.1   Example: Neutral Plane for a Rectangular Cross-section

Calculate the position of the neutral plane in a beam of rectangular cross-section.

Rectangular cross-sections are common in MEMS devices, and therefore the position of the neutral plane is often required. Figure 3.5 shows a cross-section where an arbitrary assumption has been made about the origin of coordinates (neutral plane).

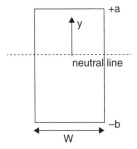

Figure 3.5   Rectangular cross-section of a beam

We assume that the origin of coordinates or location of the neutral plane is at a distance a below the top and at a distance b above the bottom. The differential of cross-section area is $dA = Wdy$:

$$\int_A y\,dA = \int_{-b}^{a} yW\,dy = \frac{1}{2}(a^2 - b^2) = 0. \tag{3.8}$$

From (3.8) it follows that $a = b$, indicating that the neutral surface is located at the mid-point of the height of the cross-section.

3.4.2   Example: Cantilever with Point Force at the Tip

We have a beam bent upwards as shown in Figure 3.6. We know that the deflection at $x = L$ is 0.5 μm and $L = 500$ μm. Calculate the value of the radius of curvature ρ_c and the value of the stress at the edges of the beam ($y = \pm h/2$) with $h = 5$ μm.

From Figure 3.6 we can write

$$\cos\theta = \frac{\rho_c - v(L)}{\rho_c}, \quad \sin\theta \simeq \frac{L}{\rho_c}$$

$$\sin\theta = \frac{\frac{2v(L)}{L}}{1 + \frac{v(L)^2}{L^2}} = 1.999 \times 10^{-3}, \quad \theta = 0.1145°$$

$$\rho_c = \frac{L}{\sin\theta} = 0.25\,\text{m}. \tag{3.9}$$

The stress relates to the position y as shown in equation (3.4):

$$\sigma_x = -E\frac{y}{\rho_c} = -\frac{130 \times 10^9}{0.25} y$$

$$\sigma_x(y = h/2) = -5.2 \times 10^{11}\,\text{Pa.} \tag{3.10}$$

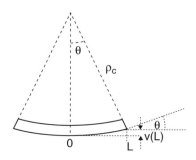

Figure 3.6   Example 3.4.2

3.5   Moment of Inertia and Bending Moment

In a beam subject to deformation there is a distribution of stress in the cross-section. Our objective is to find the relationship between the stress distribution and the bending moment. We define as a positive stress that in the direction of the positive x-axis. We also define a positive bending moment M_b as having a counterclockwise direction around the z-axis. According to Figure 3.4, we can write

$$M_b = -\int_A \sigma_x y \, dA. \tag{3.11}$$

Using equation (3.4),

$$M_b = \int_A \frac{E}{\rho_c} y^2 \, dA. \tag{3.12}$$

We define the second moment of the cross-section as the moment of inertia, I, or second moment of area:

$$I = \int_A y^2 \, dA. \tag{3.13}$$

Assuming that the radius of curvature and Young's modulus are constant, the bending moment is thus

$$M_b = \frac{EI}{\rho_c}. \tag{3.14}$$

3.5.1   Example: Moment of Inertia of a Rectangular Cross-section

Calculate the moment of inertia of a beam having a rectangular cross-section.

The bending geometry creates a moment about the z-axis. The moment of inertia is I_z and was defined in equation (3.13). The neutral plane lies in the middle of the height of the beam, as shown in Figure 3.7. Taking into account that the differential of area is $dA = W dy$, the moment of inertia I_z is calculated as

$$I_z = \int_A y^2 \, dA = \int_{-h/2}^{+h/2} W y^2 \, dy = \frac{W h^3}{12}. \tag{3.15}$$

As can be seen, the moment of inertia depends on the third power of the height of the cross-section and is linear in the width W.

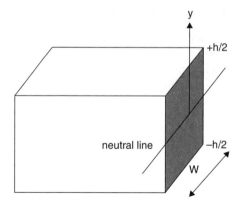

Figure 3.7 Rectangular cross-section of a beam for calculating the moment of inertia

3.6 Beam Equation

Consider the geometry of a bent beam shown in Figure 3.8. We can write [7, p. 514]

$$\tan \theta = \frac{dv}{dx}. \tag{3.16}$$

For small angles,

$$dx \approx \rho_c d\theta, \tag{3.17}$$

we have

$$\frac{1}{\rho_c} = \frac{d\theta}{dx} \tag{3.18}$$

and

$$\frac{1}{\rho_c} = \frac{d^2v}{dx^2}. \tag{3.19}$$

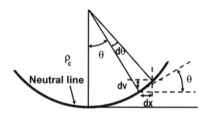

Figure 3.8 Definition of angles in neutral surface bending

Taking into account equation (3.14), the beam equation is

$$\frac{d^2v}{dx^2} = \frac{M_b}{EI}.$$ (3.20)

The solution of the beam equation subject to appropriate boundary conditions provides the deflection v as a function of the position x. The solution depends on the bending moment, which in general will be a function of the position, and on Young's modulus and moment of inertia.

3.7   End-loaded Cantilever

We consider in this section a simple case of a cantilever clamped at the left-hand end and loaded at the other end with a force F as shown in Figure 3.9.

A reaction force R and a reaction moment M_R are assumed at the left-hand end. Applying static equilibrium conditions to the whole beam:

$$\sum F = 0, \quad F - R = 0, \quad R = F;$$

$$\sum M = 0, \quad -M_R + FL = 0, \quad M_R = FL.$$ (3.21)

The result simply indicates that the reaction force R equals the point force applied and that the reaction moment cancels the counterclockwise moment created by the point force around the origin of coordinates.

We now consider the free body diagram of the section to the left of x and again apply static equilibrium conditions:

$$\sum F = 0, \quad R + Q = 0, \quad Q = -R = -F;$$

$$\sum M = 0, \quad M_b - M_R - Qx = 0, \quad M_b = F(L - x).$$ (3.22)

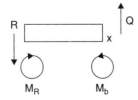

Figure 3.9   Free body diagram of a cantilever

At this point the bending moment M_b as a function of the position is known and the beam equation can be written:

$$\frac{d^2v}{dx^2} = \frac{F(L-x)}{EI}.$$ (3.23)

Integrating yields

$$\frac{dv}{dx} = \frac{F}{EI}\left(Lx - \frac{x^2}{2}\right) + A,$$ (3.24)

where A is an integration constant. Integrating again,

$$v = \frac{F}{EI}\left(L\frac{x^2}{2} - \frac{x^3}{6}\right) + Ax + B,$$ (3.25)

where B is a second integration constant. The boundary conditions that must be applied to solve for the integration constants A and B are the restrictions corresponding to the fixed end support: $v(x = 0) = 0$ and $(dv/dx)_{x=0} = 0$. It follows that $A = 0$ and $B = 0$, and

$$v = \frac{F}{EI}\left(L\frac{x^2}{2} - \frac{x^3}{6}\right).$$ (3.26)

The value of the deflection at the right-hand end of the cantilever is given by

$$v(L) = \frac{F}{EI}\left(\frac{L^3}{2} - \frac{L^3}{6}\right) = \frac{FL^3}{3EI}.$$ (3.27)

From this result we see that if the force F is positive (upwards) the deflection $v(L)$ is upwards and hence also positive.

The result can be interpreted as the constitutive relationship of a linear spring where the force and the displacement are proportional. The proportionality constant is the spring stiffness or elastic constant defined as

$$k = \frac{F}{v}.$$ (3.28)

As the force is applied at the tip, we write

$$k = \frac{F}{v(x = L)} = \frac{3EI}{L^3}.$$ (3.29)

For a rectangular cross-section of width W and thickness t, we know that $I = W^3t/12$, and hence

$$k = \frac{EW^3t}{4L^3}.$$ (3.30)

From this result it is clear that a longer beam is softer than a short one for the same cross-section.

In MEMS applications a wide variety of beams or cantilevers with different boundary conditions and loads are encountered, and the solutions can be found in specialist books such as [10, p. 189].

3.8 Equivalent Stiffness

Microstructures are often more complex than simple beams and it is very useful to derive general rules applying to the combination of different elastic supports. The rules are easy to state for linear springs where force and displacement are proportional. This is illustrated in Figure 3.10.

The parallel arrangement is shown in the upper part of the figure where two legs supporting a rigid body (shaded) are shown. The total force F is split between F_1 and F_2, but the displacement (dashed) is the same for both legs, thus $\delta_1 = \delta_2 = \delta$ and $F = F_1 + F_2$; then

$$F = F_1 + F_2 = k_1\delta_1 + k_2\delta_2 = (k_1 + k_2)\delta = k_{eq}\delta \tag{3.31}$$

and

$$k_{eq} = k_1 + k_2. \tag{3.32}$$

We can conclude that in a parallel arrangement where the displacement is the same, the equivalent stiffness is the sum of the partial stiffness values.

A series arrangement of springs is shown in the lower part of Figure 3.10. Here the applied force is the same in the two legs $F = F_1 = F_2$ but the total elongation is split between the two segments, thus

$$F = F_1 = F_2 = k_1\delta_1 = k_2\delta_2 = k_{eq}(\delta_1 + \delta_2) = k_{eq}\delta \tag{3.33}$$

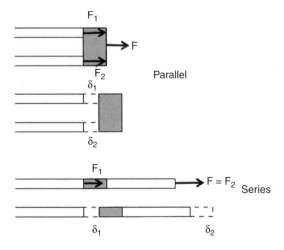

Figure 3.10   Equivalent stiffness in parallel and series arrangements

and solving yields

$$\frac{1}{k_{eq}} = \frac{1}{k_1} + \frac{1}{k_2}. \tag{3.34}$$

For a series arrangement of springs where the force is the same, the inverse of the equivalent stiffness is the sum of the inverses of the partial stiffness values.

3.9   Beam Equation for Point Load and Distributed Load

The beam equation given in equation (3.20) is written in terms of the bending moment. Sometimes it is worth writing the same equation when the load is a point or a distributed load. The conversion can be accomplished by applying force and momentum equilibrium conditions in an infinitesimal section of the beam [15, p. 201]. Let Q be a point load and q is a distributed load per unit length. Applying the force and moment equilibrium conditions, we have

$$q dx + Q + dQ - Q = 0, \quad q = -\frac{dQ}{dx}, \tag{3.35}$$

and

$$M_b + dM_b - M_b - q dx \frac{dx}{2} + (Q + dQ) dx = 0. \tag{3.36}$$

Using equations (3.35) and (3.36),

$$dM_b + \frac{dQ}{dx} \frac{dx^2}{2} + (Q + dQ) dx = 0, \quad dM_b + Q dx \simeq 0, \tag{3.37}$$

and

$$Q = -\frac{dM_b}{dx}. \tag{3.38}$$

Taking the derivative twice in equation (3.20) leads to

$$\frac{d^3 v}{dx^3} = -\frac{Q}{EI}, \quad \frac{d^4 v}{dx^4} = -\frac{q}{EI}. \tag{3.39}$$

3.10   Castigliano's Second Theorem

There are cases where the complete solution of the beam equation is not required, just the deformation or displacement at a given point. Castigliano's theorem allows a fast and simple solution. It states that the displacement δ_i colinear with a force F_i at a point of application can be calculated by the partial derivative of the total strain energy stored with respect to the

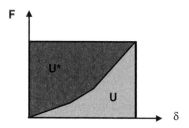

Figure 3.11   Energy and co-energy definition

force, assuming all other forces remain constant. A displacement in the direction of the force is positive. Mathematically [16],

$$\delta_i = \frac{\partial U}{\partial F_i}. \tag{3.40}$$

In some two-dimensional flexures used in MEMS there is also torsion involved, in which case the theorem can be written as

$$\theta_i = \frac{\partial U}{\partial M_{ti}}, \tag{3.41}$$

where θ_i is the rotation angle and M_{ti} is the torsion moment.

For a conservative system of a generic nonlinear spring, the force and the displacement are related as shown in Figure 3.11. The energy U is given by

$$U = \int_0^\delta F d\delta, \tag{3.42}$$

and the co-energy U^* by

$$U^* = \int_0^F \delta dF. \tag{3.43}$$

We can write

$$d(F\delta) = F d\delta + \delta dF = dU + dU^* \tag{3.44}$$

and

$$dU^* = \left(\frac{\partial U^*}{\partial F}\right)_\delta dF + \left(\frac{\partial U^*}{\partial \delta}\right)_F d\delta. \tag{3.45}$$

Finally, comparing equations (3.44) and (3.45),

$$\delta = \left(\frac{\partial U^*}{\partial F} \right)_\delta . \tag{3.46}$$

A simple particular case is a linear spring where

$$F = k\delta, \tag{3.47}$$

$$U = \int_0^\delta Fd\delta = \frac{1}{2}k\delta^2 \tag{3.48}$$

and

$$U^* = \int_0^F \delta dF = \frac{1}{2}\frac{F^2}{k}. \tag{3.49}$$

It can be seen that for a linear spring $U = U^*$, and energy and co-energy in this case can be written in different equivalent forms:

$$U = U^* = \frac{1}{2}k\delta^2 = \frac{1}{2}\frac{F^2}{k} = \frac{1}{2}F\delta. \tag{3.50}$$

3.10.1 Strain Energy in an Elastic Body Subject to Pure Bending

If we consider a differential volume inside an elastic body, and we suppose that an axial stress σ_x is applied, the differential of the strain energy dU can be written as

$$dU = \frac{1}{2}\sigma_x dydz \cdot \epsilon_x dx, \quad U = \int_V \frac{1}{2}\sigma_x \epsilon_x dxdydz. \tag{3.51}$$

We know that in pure bending,

$$\sigma_x = -\frac{M_b y}{I}, \quad \epsilon_x = \frac{\sigma_x}{E}, \tag{3.52}$$

and then

$$U = \int_L \frac{M_b^2}{2EI^2}dx \int_L y^2 dydz = \int \frac{M_b^2}{2EI}dx. \tag{3.53}$$

3.11 Flexures

Movable beams of plates are essential to the design of MEMS devices, and the flexures supporting them need to meet specifications concerning the stiffness. It is generally assumed that the flexures behave as linear springs. Some flexures have become standard.

3.11.1 Fixed–fixed Flexure

The fixed–fixed flexure for a membrane or inertial mass is shown in Figure 3.12. The aim here is to calculate the stiffness in the three directions x, y and z following [16]. Consider first the deformation in the x-direction as shown in Figure 3.12. It is assumed that, after deformation, the two end angles (at the support and at the membrane) are zero (fixed–guided end conditions). The application of Castigliano's theorem starts by isolating one of the supporting legs as shown in Figure 3.13. From Figure 3.13 the equation of moment equilibrium is

$$M_0 - M_R - F_x L = 0, \quad \text{or} \quad M_R = M_0 - F_x L \tag{3.54}$$

From the free body diagram also shown in Figure 3.13,

$$M_b - M_R - F_x y = 0 \quad \text{or} \quad M_b = M_R + F_x y = M_0 - F_x (L - y). \tag{3.55}$$

If we write $L - y = \zeta$, then

$$M_b = M_0 - F_x \zeta. \tag{3.56}$$

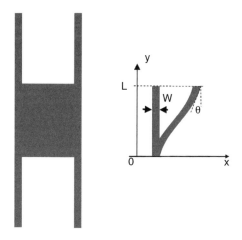

Figure 3.12   Fixed–fixed flexure

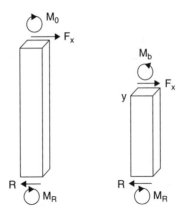

Figure 3.13 Isolated leg of a fixed-fixed flexure

The elastic energy

$$U = \int_0^L \frac{M_b^2}{2EI_z} dy = -\int_L^0 \frac{M_b^2}{2EI_z} d\zeta = \int_0^L \frac{M_b^2}{2EI_z} d\zeta, \tag{3.57}$$

where I_z is the moment of inertia about the z-axis. Taking into account that this moment is the second moment of a differential cross-section normal to the neutral surface and that the neutral surface is in the (y, z) plane (see Problem 3.2), it follows that

$$I_z = \frac{W^3 t}{12}. \tag{3.58}$$

According to Castigliano's theorem the deflection angle at the support $\theta = 0$ has to be zero for a fixed support, and hence

$$\theta = \frac{\partial U}{\partial M_0} = 0. \tag{3.59}$$

M_0 and M_b are found to be

$$M_0 = \frac{F_x L}{2}, \quad M_b = F_x \left(\frac{L}{2} - \zeta \right) \tag{3.60}$$

Substituting in equation (3.58), the value of the displacement along x can be calculated:

$$\delta_x = \frac{\partial U}{\partial F_x} = \frac{F_x L^3}{12 EI_z} \tag{3.61}$$

and

$$k_x = \frac{F_x}{\delta_x} = \frac{12 EI_z}{L^3}. \tag{3.62}$$

Exactly the same derivation can be done when the force is applied on the z-axis (out of plane), as the main change is that the moment of inertia I_x has the same expression as I_z but interchanging W and t,

$$I_x = \frac{t^3 W}{12}, \quad k_z = \frac{F_z}{\delta_z} = \frac{4EWt^3}{L^3}. \tag{3.63}$$

Finally, when the force is along the y-axis, the force is applied axially to the beam and then the stiffness is calculated straightforwardly:

$$\epsilon_y = \frac{\sigma_y}{E} = \frac{\delta_y}{L} = \frac{F_y}{tWE}, \quad k_y = \frac{F_y}{\delta_y} = \frac{EWt}{L}. \tag{3.64}$$

3.11.2 Example: Comparison of Stiffness Constants

Compare the stiffness constants of a leg in a fixed–fixed flexure made of silicon nitride, where $W = 5\ \mu m$, $t = 1\ \mu m$, $L = 200\ \mu m$ and $E = 310$ GPa.

Using equations (3.62)–(3.64), it follows that

$$k_x = \frac{F_x}{\delta_x} = \frac{12EI_z}{L^3} = 4.68\ \text{N/m}, \tag{3.65}$$

$$k_z = \frac{F_x}{\delta_x} = \frac{12EI_x}{L^3} = 0.18\ \text{N/m} \tag{3.66}$$

and

$$k_y = \frac{EWt}{L} = 7.5 \times 10^3\ \text{N/m}. \tag{3.67}$$

As can be seen, the stiffest direction of bending is along the y-axis, while the softest is in the z-direction (out of plane).

3.11.3 Example: Folded Flexure

An elastic structure often used to support a moving part is the folded flexure shown in Figure 3.14. A structure is anchored to the substrate (areas shaded black). All the rest is movable, and the movement in the direction of the x-axis is considered here. The structure has four folded beams, one of them shaded grey in the figure. Concentrating first on this grey part, and supposing that the truss on the top is totally rigid, we observe that this is similar to two fixed–guided beams connected in series: one from the anchor to the truss and the other from the truss to the movable structure. So these two beams together form one of the legs of the

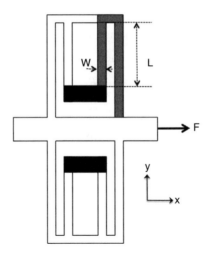

Figure 3.14 Folded flexure: areas shaded black are anchors, and one of the legs is shaded grey

flexure. If we suppose that the two beams are of the same length L, we can write, according to equation (3.62),

$$k_x = \frac{12EI_z}{L^3},$$ (3.68)

and the series combination of the two beams gives an equivalent stiffness constant,

$$k_{\text{leg}} = \frac{1}{\frac{1}{k_x} + \frac{1}{k_x}} = \frac{k_x}{2} = \frac{6EI_z}{L^3}.$$ (3.69)

The complete folded flexure has four such legs in parallel, so the equivalent stiffness is four times that of one leg,

$$k_{\text{eq}} = 4k_{\text{leg}} = \frac{24EI_z}{L^3}.$$ (3.70)

3.12 Rectangular Membrane

The two-dimensional plate equation [17] for a distributed load such as a pressure P is given by

$$\frac{\partial^4 u}{\partial x^4} + 2\frac{\partial^4 u}{\partial x^2 \partial y^2} + \frac{\partial^4 u}{\partial y^4} = \frac{P}{D},$$ (3.71)

where D is the flexure rigidity. Accurate solutions have been derived using numerical analysis for several boundary conditions in [18].

A convenient simplification of the two-dimensional plate model involves assuming that the plate is very thin compared to the other dimensions and considering the part of the membrane sufficiently far from two of the edges as a doubly supported beam with plane stress geometry as in Example 2.3.1:

$$\epsilon_x = \frac{1}{E}(\sigma_x - v\sigma_y), \quad \epsilon_y = \frac{1}{E}(\sigma_y - v\sigma_x). \tag{3.72}$$

If we suppose that the edges in the y-axis are far from the section considered, we can assume that $\epsilon_y = 0$ and $\sigma_y = -v\sigma_x$, and then

$$\epsilon_x = \frac{\sigma_x}{E}(1 - v^2), \quad \sigma_x = \epsilon_x \frac{E}{1 - v^2}. \tag{3.73}$$

Approximate solutions can be obtained using the beam equation of a beam where the Young's modulus is replaced by $E/(1 - v^2)$. If we also suppose that the cross-section is rectangular with width W and membrane thickness t, and that the moment of inertia is $I = Wt^3/12$,

$$\frac{d^2v}{dx^2} = \frac{M_b}{\frac{E}{1-v^2}I} = \frac{12M_b(1 - v^2)}{EWt^3}. \tag{3.74}$$

Now defining the flexure rigidity D as

$$D = \frac{Et^3}{12(1 - v^2)}, \tag{3.75}$$

the beam equation is simply

$$\frac{d^2v}{dx^2} = \frac{M_b}{WD}. \tag{3.76}$$

3.13 Simplified Model for a Rectangular Membrane Under Pressure

The geometry is shown in Figure 3.15, where we will consider the edges clamped. In order to simplify the situation and transform it into a one-dimensional problem, only the portion shown in Figure 3.15 will be considered, and we will assume that the behaviour is that of a doubly clamped beam of width W [19, p. 75]. We will suppose that the transversal edges of the membrane are sufficiently far from the portion selected to neglect their effect. The membrane is subject to a pressure P in the upper side. The equations giving the reactions at the supports can be written

$$R = \frac{PLW}{2} \tag{3.77}$$

The free body diagram from $x = 0$ to x is also drawn in Figure 3.15. The shear force V is calculated as

$$R - V - PWx = 0, \quad V = \frac{PWL}{2} - PWx. \tag{3.78}$$

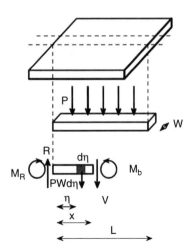

Figure 3.15   Rectangular membrane: quasi-1D model

Using an auxiliary variable η and taking moments around $x = 0$,

$$M_b - M_R - Vx - \int_0^x PW\eta \, d\eta = 0 \quad \text{and} \quad M_b - M_R - Vx - PW\frac{x^2}{2} = 0. \qquad (3.79)$$

Substituting V,

$$M_b - M_R - \frac{PWL}{2}x + PWx^2 - PW\frac{x^2}{2} = 0 \qquad (3.80)$$

and

$$M_b = M_R + \frac{PWL}{2}x - PW\frac{x^2}{2}. \qquad (3.81)$$

The value of M_R is still unknown. Integrating the beam equation,

$$\frac{d^2v}{dx^2} = \frac{M_b}{EI} = \frac{1}{EI}\left(M_R + \frac{PWL}{2}x - PW\frac{x^2}{2}\right). \qquad (3.82)$$

Integrating twice,

$$v = \frac{1}{EI}\left(M_R\frac{x^2}{2} + \frac{PWL}{2}\frac{x^3}{6} - PW\frac{x^4}{24}\right) + Ax + B. \qquad (3.83)$$

Setting the clamped boundary conditions, $v(x = 0) = 0$ and $dv/dx(x = 0) = 0$, leads to $A = B = 0$. We also apply the boundary condition that $v(L) = 0$ at $x = L$, and then solve for M_R:

$$v(L) = \frac{1}{EI}\left(M_R\frac{L^2}{2} + \frac{PWL}{2}\frac{L^3}{6} - PW\frac{L^4}{24}\right) = 0 \quad \text{and} \quad M_R = -\frac{PWL^2}{12}. \qquad (3.84)$$

Then the deflection is

$$v = -\frac{PW}{24EI}x^2(x-L)^2.$$

(3.85)

Substituting the equation for the moment of inertia for a rectangular cross-section,

$$v(x) = \frac{P}{2Et^3}x^2(x-L)^2.$$

(3.86)

This result, which has been found for a portion of the membrane, can be extended to an approximate solution for the full rectangular membrane by substituting Young's modulus in equation (3.86) by $E/(1-v^2)$. Similarly, the stress as a function of the position x can be calculated as

$$\sigma_x = \frac{M_b y}{I}.$$

(3.87)

The maximum stress will be at the two surfaces of the membrane, namely at $y = \pm h/2$:

$$\sigma_x\left(\pm\frac{h}{2}\right) = \frac{PWt}{2I}\left(-\frac{L^2}{12} + \frac{L}{2}x - \frac{x^2}{2}\right).$$

(3.88)

3.13.1 Example: Thin Membrane Subject to Pressure

A membrane 400 μm square and 1 μm thick is subject to a pressure of 1×10^4 Pa. For the calculation we consider a portion of the membrane of width $W = 10\,\mu m$. Young's modulus is 160 GPa. Calculate the deflection $v(x)$ of this portion of the membrane as a function of the position x. Calculate and plot the values of the stress at the top and bottom surfaces of the membrane as a function of the position x.

The deflection is calculated from equation (3.86),

$$v(x) = \frac{P}{2Et^3}x^2(x-L)^2 = \frac{1\times10^4}{2\times160\times10^9(1\times10^{-6})^3}x^2(x-400\times10^{-6})^2.$$

(3.89)

The stress is a maximum at the two membrane surfaces,

$$\sigma_x\left(\pm\frac{h}{2}\right) = \frac{PWt}{2I}\left(-\frac{L^2}{12} + \frac{L}{2}x - \frac{x^2}{2}\right)$$

(3.90)

$$= \frac{1\times10^4\times10\times10^{-6}\times10^{-6}}{\frac{2\times10^{-6}\times1\times10^{-18}}{12}}\left(-\frac{(400\times10^6)^2}{12} + \frac{400\times10^{-6}}{2}x - \frac{x^2}{2}\right),$$

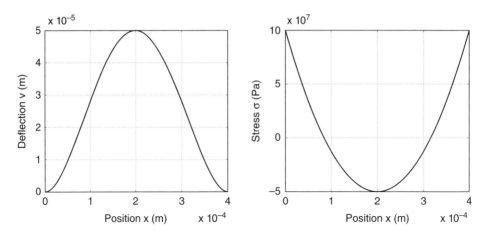

Figure 3.16 Distribution of the deflection and of stress

and

$$\sigma_x \left(\pm \frac{h}{2} \right) = 6 \times 10^{17} \left(0.013 + 2 \times 10^{-4} x - \frac{x^2}{2} \right) \text{ Pa.} \tag{3.91}$$

The results are shown in Figure 3.16, where it can be seen that the maximum value of the stress is located at the edges of the membrane, and is twice the stress at the membrane centre.

3.14 Edge-clamped Circular Membrane

In a circular membrane there is radial symmetry and at any point there are two strains: radial and tangential. In the radial direction a similar equation to that for the x-axis in a cartesian geometry as in equation (3.2) can be written,

$$\epsilon_r = -\frac{y}{\rho_c} = -y \frac{d^2 v}{dr^2}. \tag{3.92}$$

In the tangential direction, following a similar derivation to that in [20] and as can be seen in Figure 3.17, the length of the upper circle is $2\pi(r - dr)$ and that of the bottom circle is $2\pi r$. As $dr = y\theta$,

$$\epsilon_t = \frac{2\pi(r - dr) - 2\pi r}{2\pi r} = -\frac{dr}{r} = -\frac{y\theta}{r} = -\frac{y}{r} \frac{dv}{dr}. \tag{3.93}$$

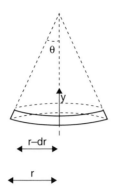

Figure 3.17 Geometry to calculate the tangential strain

The negative sign is consistent with the compressive side considered in Figure 3.17. The corresponding radial and tangential stresses are given by

$$\sigma_r = \frac{E}{1-v^2}(\epsilon_r + v\epsilon_t) = -\frac{Ey}{1-v^2}\left(\frac{d^2v}{dr^2} + \frac{v}{r}\frac{dv}{dr}\right),$$ (3.94)

$$\sigma_t = \frac{E}{1-v^2}(\epsilon_t + v\epsilon_r) = -\frac{Ey}{1-v^2}\left(\frac{1}{r}\frac{dv}{dr} + v\frac{d^2v}{dr^2}\right).$$ (3.95)

To calculate the deflection, the two-dimensional plate equation (3.71) can be written in polar coordinates as

$$\frac{1}{r}\frac{\partial}{\partial r}\left(\left(r\frac{\partial}{\partial r}\right)\left(\frac{1}{r}\frac{\partial}{\partial r}\left(r\frac{\partial}{\partial r}\right)\right)\right)v = \frac{P}{D}.$$ (3.96)

The first integration of equation (3.96) gives

$$\frac{\partial}{\partial r}\left(\frac{1}{r}\frac{\partial}{\partial r}\left(r\frac{\partial}{\partial r}\right)\right)v = \frac{Pr}{2D}$$ (3.97)

The second integration

$$\frac{1}{r}\frac{\partial}{\partial r}\left(r\frac{\partial}{\partial r}\right)v = \frac{Pr^2}{4D} + A.$$ (3.98)

The third integration

$$r\frac{\partial}{\partial r}v = \frac{Pr^3}{16D} + A\frac{r^2}{2} + B.$$ (3.99)

The fourth integration

$$v = \frac{Pr^4}{64D} + A\frac{r^2}{4} + B\ln r + C.$$ (3.100)

The three integration constants A, B and C are calculated from the boundary conditions. If we assume that the membrane edges are clamped, and that the slope at the centre is zero, it follows that

$$A = -\frac{PR^2}{8D}, \quad B = 0, \quad C = \frac{PR^4}{64D}, \tag{3.101}$$

and then

$$v = \frac{P}{64D}(r^2 - R^2)^2. \tag{3.102}$$

Once the function $v(r)$ is known the stresses can be calculated using equation (3.102) together with equations (3.94) and (3.95),

$$\frac{dv}{dr} = \frac{P}{64D}4r(r^2 - R^2), \quad \frac{d^2v}{dr^2} = \frac{P}{64D}(12r^2 - 4R^2). \tag{3.103}$$

Using equation (3.75) for the flexure rigidity D and setting $y = t/2$, where t is the membrane thickness, we obtain the maximum value of the stress at a given r,

$$\sigma_r(y = t/2) = -\frac{PR^2}{8t^2}\left(\frac{3r^2}{R^2} - 1 + v\left(\frac{r^2}{R^2} - 1\right)\right). \tag{3.104}$$

The maximum value is located at the edge,

$$\sigma_{r\,max} = -\frac{3PR^2}{4t^2}. \tag{3.105}$$

Similarly, the tangential stress component can be calculated to give

$$\sigma_t = -\frac{3PR^2}{8t^2}\left(\frac{3vr^2}{R^2} - v + \frac{r^2}{R^2} - 1\right), \tag{3.106}$$

and the maximum value of the tangential stress occurs at the centre, $r = 0$,

$$\sigma_{t\,max} = \frac{3PR^2}{8t^2}(1 + v). \tag{3.107}$$

The value of this maximum tangential stress is numerically equal to the value of the radial stress at the centre, $\sigma_{t\,max} = \sigma_r(r = 0)$.

Problems

3.1 Calculate the position of the neutral surface in a beam having a trapezoidal cross-section as shown in Figure 3.18.

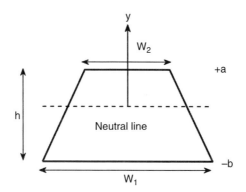

Figure 3.18   Problem 3.1

3.2   Calculate the moment of inertia for the geometries depicted in Figure 3.19. Assume a generic volume of rectangular cross-sections in all dimensions as shown in Figure 3.20.

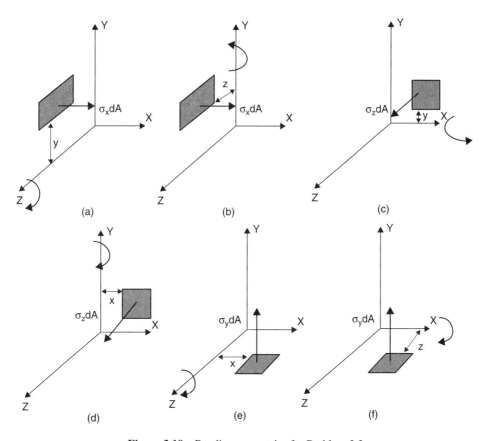

Figure 3.19   Bending geometries for Problem 3.2

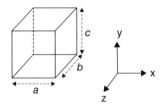

Figure 3.20   Dimensions of a generic prism for Problem 3.2

3.3 We have a cantilever that during fabrication has accumulated a residual stress causing the upper part to be under expansion and the lower part under compression. The cross-section is rectangular, and we know that the neutral plane lies in the middle. The thickness of the cantilever is $h = 2\,\mu m$. We also know that we are in pure bending and that the value of the stress at the upper surface $\sigma_0(y = h/2) = 20$ MPa, $W = 10\,\mu m$, $E = 160$ GPa. Draw the stress distribution as a function of the value of the position y inside the cantilever, assuming it is linear, and write the function $\sigma(y)$. Calculate the value of bending moment, M_b, caused by the distribution of stress at the section considered. Is the bending moment clockwise or counterclockwise? Calculate the value of the radius of curvature ρ_c.

3.4 A beam is supported at both ends by fixed supports and has a point load F at the middle $x = L/2$, as shown in Figure 3.21. Calculate the reactions at the supports, and the deflection at any point.

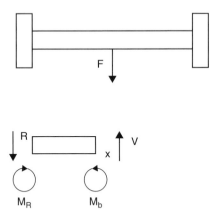

Figure 3.21   Problem 3.4

3.5 Using the results found in Problem 3.4, if the load force is $F = 0.1\,\mu N$ located at the middle point as shown in Figure C.5, find the position where the deflection is a maximum and calculate its value. We know that the cantilever has length $L = 1000\,\mu m$, width $W = 10\,\mu m$ and thickness $h = 1\,\mu m$. The cantilever is made of a material having a Young's modulus of $E = 160$ GPa.

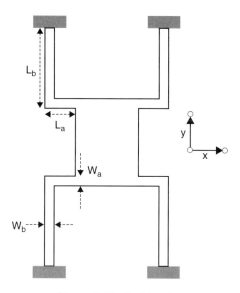

Figure 3.22   Problem 3.6

3.6   The structure of a crab leg flexure is shown in Figure 3.22. It is frequently used in MEMS devices, to support either laterally or vertically moving structures. Calculate the equation for the stiffness along the *x*-axis.

3.7   Cantilevers are used commercially for AFM force spectroscopy. In [21] dimensions of commercial cantilevers are collected and can also be found in vendor information. Typical dimensions are length $L = 203.8\,\mu$m, width $W = 20.38\,\mu$m and thickness $t = 0.55\,\mu$m. Cantilevers are made of silicon nitride. The quoted stiffness constant is 20 pN/nm. Calculate the theoretical value of the stiffness constant assuming a fixed end cantilever.

3.8   For a cantilever of the same dimensions as in Problem 3.7, we would like to calculate the distance at which a CCD camera has to be placed in order to get a displacement of $9\,\mu$m of a reflected laser beam at the CCD plane for a force applied of 10 pN.

3.9   Find the position of the neutral plane of a cantilever formed by a stack of two films of different materials, as shown in Figure 3.23. Calculate the value of b for $t_1 = 1\,\mu$m, $t_2 = 2\,\mu$m, $E_1 = 385$ GPa and $E_2 = 130$ GPa.

3.10   We know that the differential equation governing the deflection of a beam having an axial tensile stress is the Euler–Bernoulli equation given by

$$EI\frac{d^4v}{dx^4} - \sigma_0 Wh\frac{d^2y}{dx^2} = q, \tag{3.108}$$

as given in [15, p. 229] and [16]. Write Matlab code to solve this differential equation with clamped beam boundary conditions at both ends. The beam has a length of $L = 200\,\mu$m, width $W = 10\,\mu$m and thickness $h = 2\,\mu$m. Young's modulus is $E = 130$

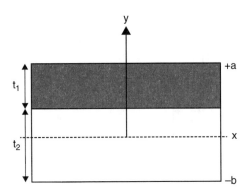

Figure 3.23   Problem 3.9

GPa. Calculate the value of q required for the beam to deflect 20% of the thickness with stress $\sigma_0 = 0$. Calculate the deflection for the same value of q found in the previous point and for three values of σ_0.

3.11 Find the definition of the centroid of an area and show that in a rectangular cross-section the neutral line passes through the centroid.

3.12 We have a stack of two materials of different thickness and Young's modulus E_1 and E_2, respectively. Show that the stack of two materials can be treated as a single material having two different widths. This is known as the 'equivalent width' method for finding the centroid of the stack.

3.13 Using the equivalent width method, find the centroid of a stack of two layers of different materials with different thickness and Young's modulus.

4

Piezoresistance and Piezoelectricity

This chapter covers the basics of piezoresistive and piezoelectric effects, both relevant in the technology of MEMS applications. The piezoresistive effect is the stress-driven change in material resistivity. Silicon shows a significant piezoresistive effect, and silicon resistors are widely used in deformation or stress sensing. Piezoelectricity denotes the interaction of the electrical and mechanical domains: mechanical stress induces electrical displacement and, conversely, an electric field induces deformation. Both, sensors and actuators are made of piezoelectric materials. Electrical resistance is first defined, and the relative change in the value of a resistor due to stress allows the definition of the gauge factor. The coordinate rotation properties introduced in Chapter 2 are extended here to find the values of the piezoresistive coefficients in a rotated coordinate system. The results are then applied to thin-film silicon resistors and piezoresistive accelerometers using cantilevers and pressure sensors using membranes. Finally, direct and inverse piezoelectric effects are discussed and their application to a voltage generator described.

4.1 Electrical Resistance

The electrical resistance of a given conductive body is defined as

$$R = \rho \frac{L}{A},\tag{4.1}$$

where ρ is the material resistivity (in units of $\Omega \cdot m$),[1] L is the length and A the cross-section normal to the direction of the electrical current, as shown in Figure 4.1.

For a width W and thickness t, the resistance is given by

$$R = \rho \frac{L}{Wt} = \frac{\rho}{t} \frac{L}{W}.\tag{4.2}$$

---

[1] In this chapter the letter ρ denotes the electrical resistivity and should not be confused with ρ_c used for the radius of curvature in Chapter 3.

Understanding MEMS: Principles and Applications, First Edition. Luis Castañer.
© 2016 John Wiley & Sons, Ltd. Published 2016 by John Wiley & Sons, Ltd.
Companion Website: www.wiley.com/go/castaner/understandingmems

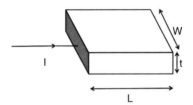

Figure 4.1   Resistance of a prismatic body

In microelectronics most of the measurements of the electrical properties of the different layers are performed from the upper side. Resistivity measurement is done using four collinear point probes where a current I is injected between the two more external probes and a voltage V is measured between the two middle probes. The ratio V/I is related to the sheet resistance $R^\square$ defined as the ratio ρ/t [22]:

$$R^\square = \frac{\rho}{t} = \frac{\pi}{\ln 2} \frac{V}{I}. \tag{4.3}$$

The resistance can be written as

$$R = R^\square \frac{L}{W}. \tag{4.4}$$

The sheet resistance has units of Ω but is generally given in $\Omega/\square$ to indicate that it is the resistance of a square. This a very useful formulation as normally the sheet resistance value is known, and when $L = W$, equation (4.4) shows that the resistance value equals that of the sheet resistance, $R = R^\square$. For rectangular shapes, the ratio L/W is simply the number of squares that can be drawn. The resistance layout often includes corners, to reduce the area budget, and connecting packet assembler/disassemblers (PADs). A square that includes a corner or a PAD is not accounted for as a full square but as a fraction.

4.1.1   Example: Resistance Value

Given a material having a sheet resistance of $15\,\Omega/\square$, we would like to build a $400\,\Omega$ resistor. The minimum width for the layout is $2\,\mu m$ and the lateral separation between two active layers should also be $2\,\mu m$. Draw the resistor and calculate the total area budget. Assume that a connecting PAD accounts for 0.14 squares and a corner for 0.55 squares.

In order to reduce the resistor size we choose $2\,\mu m$ as the resistor width. According to equation (4.4),

$$400 = 15\frac{L}{W}, \quad \text{so} \quad \frac{L}{W} = 26.67. \tag{4.5}$$

In the layout shown in Figure 4.2, there are six corners in addition to the two end PADs and 23 full squares, totalling $23 + 6 \times 0.55 + 2 \times 0.14 = 26.58$ squares.

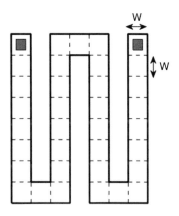

Figure 4.2 Layout of the resistor

4.2 One-dimensional Piezoresistance Model

Piezoresistance denotes the property of some materials of experiencing resistivity change when subject to stress. It is an anisotropic property and hence depends upon the relative orientation of the stress and crystal plane. Let us first consider the one-dimensional geometry shown Figure 4.3. If the rod has length L and diameter $2r$ and is made of a material with a resistivity ρ, the resistance between the two ends is given by

$$R = \rho \frac{L}{\pi r^2}. \tag{4.6}$$

When a uniaxial stress is applied, the rod deforms to a length $L + dL$ and to a radius $r - dr$. If the material resistivity experiences a change to $\rho + d\rho$, the value of the resistance change dR is given by

$$dR = \frac{L}{\pi r^2} d\rho + \frac{\rho}{\pi r^2} dL - \frac{\rho L}{\pi} \frac{2}{r^3} dr. \tag{4.7}$$

The relative change in the resistance value is

$$\frac{dR}{R} = \frac{d\rho}{\rho} + \frac{dL}{L} - 2\frac{dr}{r}, \tag{4.8}$$

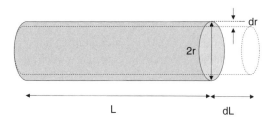

Figure 4.3 One-dimensional rod stressed

where it can be seen that there are three components of the resistance change: changes in the longitudinal and transversal dimensions and a change in resistivity. In most metals the change in resistivity is negligible, and this is why strain gauges made of metal have their sensitivity to stress limited to dimensional changes, whereas in some semiconductors the change in resistivity is the major component, generally with larger values than the other two. Taking into account that the transversal and longitudinal strains, S_t and S_l, are related by Poisson's ratio v,

$$v = -\frac{S_t}{S_l} = -\frac{dr/r}{dL/L} \quad \text{and} \quad \frac{dR}{R} = (1+2v)\frac{dL}{L} + \frac{d\rho}{\rho}. \tag{4.9}$$

The relative change in resistivity is related to the stress by means of a piezoresistance coefficient, denoted π:

$$\Delta = \frac{d\rho}{\rho} = \pi T. \tag{4.10}$$

As the stress, T, and strain, S, are related by the elastic Young's modulus E,[2]

$$T = ES \tag{4.11}$$

the relative change in resistance is

$$\frac{dR}{R} = (1 + 2v + \pi E)\,S. \tag{4.12}$$

The quantity $1 + 2v + \pi E$ is called the gauge factor (GF). In metals the geometric term predominates, while in semiconductors the piezoresistance term predominates.

4.2.1   Example: Gauge Factors

Compare the gauge factor of a piezoresistor made of aluminum with a similar resistor made of p-type silicon having the stress applied in the (110) direction. The piezoresistance coefficient for silicon in that direction is $\pi_{110} = 71.8 \times 10^{-11}\ Pa^{-1}$, Young's modulus is $E_{110} = 168.7\ GPa$ and Poisson's ratio $v_{110} = 0.0054$ (see Problem 2.3).

The aluminum has Poisson's ratio $v_{Al} = 0.33$. As aluminum is not a piezoresistive material, the gauge factor is simply given by

$$GF_{Al} = 1 + 2v_{Al} = 1 + 2 \times 0.33 = 1.66. \tag{4.13}$$

---

[2] In this chapter E is Young's modulus and $\mathbf{E}$ is the electric field intensity.

The p-type silicon in the (110) direction has gauge factor

$$GF_{110} = 1 + 2\nu_{110} + \pi_{110}E_{110} = 1 + 2 \times 0.0054 + 71.8 \ 10^{-11} \times 168.7 \times 10^9 = 121.12.$$

$$(4.14)$$

As can be seen, the piezoresistive effect greatly enhances the GF in silicon compared to aluminum.

4.3 Piezoresistance in Anisotropic Materials

Ohm's law relates the current density $\vec{J}$ with the electric field $\vec{E}$ by means of the electrical resistivity ρ:

$$\vec{E} = \rho \vec{J}. \tag{4.15}$$

More generally, the resistivity is a second-rank tensor relating the components of the electric field intensity to the components of the electric current density:

$$\begin{pmatrix} \mathbf{E}_x \\ \mathbf{E}_y \\ \mathbf{E}_z \end{pmatrix} = \begin{pmatrix} \rho_{11} & \rho_{12} & \rho_{13} \\ \rho_{21} & \rho_{22} & \rho_{23} \\ \rho_{31} & \rho_{32} & \rho_{33} \end{pmatrix} \begin{pmatrix} J_x \\ J_y \\ J_z \end{pmatrix}. \tag{4.16}$$

The resistivity tensor is symmetric:

$$\begin{pmatrix} \mathbf{E}_x \\ \mathbf{E}_y \\ \mathbf{E}_z \end{pmatrix} = \begin{pmatrix} \rho_1 & \rho_6 & \rho_5 \\ \rho_6 & \rho_2 & \rho_4 \\ \rho_5 & \rho_4 & \rho_3 \end{pmatrix} \begin{pmatrix} J_x \\ J_y \\ J_z \end{pmatrix}. \tag{4.17}$$

Of course, in an isotropic material,

$$\begin{pmatrix} \mathbf{E}_x \\ \mathbf{E}_y \\ \mathbf{E}_z \end{pmatrix} = \begin{pmatrix} \rho_0 & 0 & 0 \\ 0 & \rho_0 & 0 \\ 0 & 0 & \rho_0 \end{pmatrix} \begin{pmatrix} J_x \\ J_y \\ J_z \end{pmatrix}, \tag{4.18}$$

the resistivity tensor reduces to a diagonal matrix. Let us extend our definition of incremental resistivity change in equation (4.12) as

$$\Delta_i = \frac{\rho_i - \rho_0}{\rho_0}, \ i = 1, 2, 3, \quad \text{and} \quad \Delta_j = \frac{\rho_j}{\rho_0}, \ j = 4, 5, 6. \tag{4.19}$$

There are only six different components of Δ and six different components of the stress, and for single-crystal silicon, due to symmetry, there are only three independent values of the piezoresistance coefficients: π_{11}, π_{12} and π_{44}.

The piezoresistive tensor is

$$\begin{pmatrix} \Delta_1 \\ \Delta_2 \\ \Delta_3 \\ \Delta_4 \\ \Delta_5 \\ \Delta_6 \end{pmatrix} = \begin{pmatrix} \pi_{11} & \pi_{12} & \pi_{12} & 0 & 0 & 0 \\ \pi_{12} & \pi_{11} & \pi_{12} & 0 & 0 & 0 \\ \pi_{12} & \pi_{12} & \pi_{11} & 0 & 0 & 0 \\ 0 & 0 & 0 & \pi_{44} & 0 & 0 \\ 0 & 0 & 0 & 0 & \pi_{44} & 0 \\ 0 & 0 & 0 & 0 & 0 & \pi_{44} \end{pmatrix} \begin{pmatrix} T_1 \\ T_2 \\ T_3 \\ T_4 \\ T_5 \\ T_6 \end{pmatrix}, \tag{4.20}$$

which we can write as

$$(\Delta) = (\Pi)(T). \tag{4.21}$$

Table 4.1 gives typical values for the piezoresistive coefficients for silicon [23, 24] and silicon carbide [25].

Table 4.1   Piezoresistive coefficients, (Pa^{-1})

	π_{11}	π_{12}	π_{44}
Silicon p-type [23]	6.6×10^{-11}	-1.1×10^{-11}	138.1×10^{-11}
Silicon n-type [24]	-102.2×10^{-11}	53.4×10^{-11}	-13.6×10^{-11}
SiC [25]	1.5×10^{-11}	-1.4×10^{-11}	18.1×10^{-11}

Polysilicon can be modelled as a series of single-crystalline silicon grains separated by grain boundaries. An estimate of longitudinal and tangential piezoresistive coefficients averaging all orientations is given in [26]:

$$\pi_l = \pi_{11} - 0.4(\pi_{11} - \pi_{12} - \pi_{44}), \quad \pi_t = \pi_{12} + 0.133(\pi_{11} - \pi_{12} - \pi_{44}). \tag{4.22}$$

4.4   Orthogonal Transformation of Ohm's Law

In Chapter 2 we described the transformation of the stiffness and compliance matrices subject to an orthogonal transformation of coordinates. Here we can proceed similarly, taking into account that Ohm's law in the unrotated coordinate system is

$$\vec{E} = (\rho)\vec{J}, \tag{4.23}$$

and after rotation,

$$\vec{E'} = (\rho')\vec{J'}. \tag{4.24}$$

If the rotation matrix is (R), then

$$\vec{E'} = (R)\vec{E}, \quad \vec{J'} = (R)\vec{J}. \tag{4.25}$$

Substituting into equation (4.24),

$$(R)\vec{E} = (\rho')(R)\vec{J}. \tag{4.26}$$

Multiplying equation (4.26)on the left by $(R)^{-1}$,

$$(R)^{-1}(R)\vec{E} = (R)^{-1}(\rho')(R)\vec{J} \tag{4.27}$$

hence

$$\vec{E} = (R)^{-1}(\rho')(R)\vec{J}. \tag{4.28}$$

Comparing equation (4.23) to equation (4.28),

$$(\rho) = (R)^{-1}(\rho')(R) \quad \text{and} \quad (\rho') = (R)(\rho)(R)^{-1}, \tag{4.29}$$

or

$$(\rho') = \begin{pmatrix} l_1 & m_1 & n_1 \\ l_2 & m_2 & n_2 \\ l_3 & m_3 & n_3 \end{pmatrix} \cdot \begin{pmatrix} \rho_1 & \rho_6 & \rho_5 \\ \rho_6 & \rho_2 & \rho_4 \\ \rho_5 & \rho_4 & \rho_3 \end{pmatrix} \cdot \begin{pmatrix} l_1 & l_2 & l_3 \\ m_1 & m_2 & m_3 \\ n_1 & n_2 & n_3 \end{pmatrix}. \tag{4.30}$$

Taking into account that the rotation matrix satisfies the relationship $(R)^{-1} = R^T$, equation (4.30) is the same transformation equation as we used for the stress in equation (2.79), and

$$\begin{pmatrix} \rho_1' \\ \rho_2' \\ \rho_3' \\ \rho_4' \\ \rho_5' \\ \rho_6' \end{pmatrix} = \begin{pmatrix} l_1^2 & m_1^2 & n_1^2 & 2m_1n_1 & 2n_1l_1 & 2l_1m_1 \\ l_2^2 & m_2^2 & n_2^2 & 2m_2n_2 & 2n_2l_2 & 2l_2m_2 \\ l_3^2 & m_3^2 & n_3^2 & 2m_3n_3 & 2n_3l_3 & 2l_3m_3 \\ l_2l_3 & m_2m_3 & n_2n_3 & m_2n_3+m_3n_2 & n_2l_3+n_3l_2 & m_2l_3+m_3l_2 \\ l_3l_1 & m_3m_1 & n_3n_1 & m_3n_1+m_1n_3 & n_3l_1+n_1l_3 & m_3l_1+m_1l_3 \\ l_1l_2 & m_1m_2 & n_1n_2 & m_1n_2+m_2n_1 & n_1l_2+n_2l_1 & m_1l_2+m_2l_1 \end{pmatrix} \cdot \begin{pmatrix} \rho_1 \\ \rho_2 \\ \rho_3 \\ \rho_4 \\ \rho_5 \\ \rho_6 \end{pmatrix}, \tag{4.31}$$

or in short form,

$$(\rho') = (N) \cdot (\rho). \tag{4.32}$$

4.5 Piezoresistance Coefficients Transformation

The general transformation of coordinates of the six resistivity components shown in equation (4.32) can be extended to the transformation of the six components of stress and the six components of the incremental changes in the resistivity Δ:

$$(T') = (N)(T), \quad (\Delta') = (N)(\Delta), \tag{4.33}$$

and

$$(\Delta') = (\pi')(T').$$ (4.34)

Substituting equation (4.33),

$$(N) \cdot (\Delta) = (\pi') \cdot (N) \cdot (T).$$ (4.35)

Multiplying both sides by $(N)^{-1}$,

$$(\Delta) = (N)^{-1} \cdot (\pi') \cdot (N) \cdot (T),$$ (4.36)
$$(\pi') = (N) \cdot (\pi) \cdot (N)^{-1}.$$ (4.37)

If we let

$$(M) = (N)^{-1},$$ (4.38)

equation (4.37) can be expanded as

$$\pi'_{ij} = \sum_{k,l=1}^{6} N_{ik}\pi_{kl}M_{lj}.$$ (4.39)

Once the rotation matrix is known, the piezoresistive coefficients can be calculated from equation (4.39).

4.5.1 Example: Calculation of Rotated Piezoresistive Components π'_{11}, π'_{12} and π'_{16} for unit axes X′ [110], Y′ [$\bar{1}$10] and Z′ [001]

For a silicon wafer (100), the plane of the wafer is one of the planes equivalent to (100), so we can still use our definition of the main crystallographic axes. The set of axes X′ [110], Y′ [$\bar{1}$10] and Z′ [001] is an orthogonal set and suitable for many MEMS applications, as the flat is readily identified. The two coordinate systems are shown in Figure 4.4.

The Z- and Z′-axes coincide, and the angle between the X- and X′-axes is 45°, as is the angle between the the Y- and Y′-axes. The rotation matrix in this case has already been calculated in equation (2.78), reproduced here:

$$\begin{pmatrix} l_1 & m_1 & n_1 \\ l_2 & m_2 & n_2 \\ l_3 & m_3 & n_3 \end{pmatrix} = \begin{pmatrix} \frac{1}{\sqrt{2}} & \frac{1}{\sqrt{2}} & 0 \\ -\frac{1}{\sqrt{2}} & \frac{1}{\sqrt{2}} & 0 \\ 0 & 0 & 1 \end{pmatrix}.$$ (4.40)

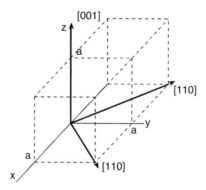

Figure 4.4 Axes of a silicon wafer (100)

In this case the matrix (N) is given by

$$(N) = \begin{pmatrix} 0.5 & 0.5 & 0 & 0 & 0 & 1 \\ 0.5 & 0.5 & 0 & 0 & 0 & -1 \\ 0 & 0 & 1 & 0 & 0 & 0 \\ 0 & 0 & 0 & \sqrt{0.5} & -\sqrt{0.5} & 0 \\ 0 & 0 & 0 & \sqrt{0.5} & \sqrt{0.5} & 0 \\ -0.5 & 0.5 & 0 & 0 & 0 & 0 \end{pmatrix},$$ (4.41)

and its inverse, (M),

$$(M) = (N)^{-1} = \begin{pmatrix} 0.5 & 0.5 & 0 & 0 & 0 & -1 \\ 0.5 & 0.5 & 0 & 0 & 0 & 1 \\ 0 & 0 & 1 & 0 & 0 & 0 \\ 0 & 0 & 0 & \sqrt{0.5} & \sqrt{0.5} & 0 \\ 0 & 0 & 0 & -\sqrt{0.5} & \sqrt{0.5} & 0 \\ 0.5 & -0.5 & 0 & 0 & 0 & 0 \end{pmatrix}.$$ (4.42)

Matlab code can be written to calculate the result of equation (4.39):

```
Calculation of rotated piezoresistive coefficient Piprime(i,j)
i=1
j=1
piprime=0;
for k=1:6
    for l=1:6
        piprime=piprime+N(i,k)*Pi(k,l)*M(l,j);
    end
end
solution=piprime
```

Using this code for $i = 1$ and $j = 1$ gives π'_{11}, for $i = 1$ and $j = 2$ gives π'_{12}, and for $i = 1$ and $j = 6$ gives π'_{16}. The results are as follows:

$$\pi'_{11} = 7.18 \times 10^{-10}\,\mathrm{Pa}^{-1}, \quad \pi'_{12} = 6.63 \times 10^{-10}\,\mathrm{Pa}^{-1}, \quad \pi'_{16} = 7.70 \times 10^{-11}\,\mathrm{Pa}^{-1}.$$

$$(4.43)$$

4.5.2 Analytical Expressions for Some Rotated Piezoresistive Components

Analytical expressions can be derived for the rotated components of the piezoresistive matrix. In this chapter we will use the components π'_{11}, π'_{12}, π'_{16}, π'_{61}, π'_{62} and π'_{66}. The analytical expressions can be found in [19, p. 218]:

$$
\begin{aligned}
\pi'_{11} &= \pi_{11} - 2\pi_0 \left(l_1^2 m_1^2 + l_1^2 n_1^2 + m_1^2 n_1^2 \right), \\
\pi'_{12} &= \pi_{12} + \pi_0 \left(l_1^2 l_2^2 + m_1^2 m_2^2 + n_1^2 n_2^2 \right), \\
\pi'_{16} &= 2\pi_0 \left(l_1^3 l_2 + m_1^3 m_2 + n_1^3 n_2 \right), \\
\pi'_{61} &= \pi_0 \left(l_1^3 l_2 + m_1^3 m_2 + n_1^3 n_2 \right), \\
\pi'_{62} &= \pi_0 \left(l_1 l_2^3 + m_1 m_2^3 + n_1 n_2^3 \right), \\
\pi'_{66} &= \pi_{44} - 2\pi_0 \left(l_1^2 l_2^2 + m_1^2 m_2^2 \right),
\end{aligned}
$$

$$(4.44)$$

where

$$\pi_0 = \pi_{11} - \pi_{12} - \pi_{44}.$$

$$(4.45)$$

See Problem 4.1 for the derivation of the analytical expression for π'_{11}.

4.6 Two-dimensional Piezoresistors

Many applications of piezoresistive sensing are based on microstructures in which one of the dimensions (thickness) is much smaller than the other two (length and width). The current only flows in one of the dimensions of the resistor. This is illustrated in Figure 4.5. The coordinate system for the piezoresistance is (X', Y', Z'), whereas the crystallographic silicon crystal axes are (X, Y, Z). We neglect the Z'-components of the electric field ($E'_z = 0$) and of the current density ($J'_z = 0$), and also neglect the Y'-component of the current density ($J'_y = 0$). The current I'_x flows along the X' axis as shown in the figure. This allows us to write a two-dimensional Ohm's law,

$$\begin{pmatrix} E'_x \\ E'_y \end{pmatrix} = \begin{pmatrix} \rho'_1 & \rho'_6 \\ \rho'_6 & \rho'_2 \end{pmatrix} \begin{pmatrix} J'_x \\ 0 \end{pmatrix}.$$

$$(4.46)$$

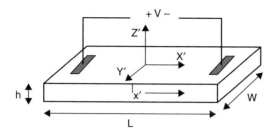

Figure 4.5 Geometry of a two-dimensional piezoresistance

From 4.46, for the X'-component of the electric field we have

$$\mathbf{E}'_x = \rho'_1 J'_x \tag{4.47}$$

Taking into account that the applied voltage V creates an electric field E'_x given by

$$\mathbf{E}'_x = \frac{V}{L}, \tag{4.48}$$

the resistance is

$$R = \frac{V}{I'_x} = \rho'_1 \frac{L}{Wh}. \tag{4.49}$$

As the total current I'_x is

$$I'_x = J'_x Wh, \tag{4.50}$$

the electrical resistance is

$$R = \frac{V}{I'_x} = \rho'_1 \frac{L}{Wh}. \tag{4.51}$$

If the value of the resistance, in absence of stress, is R_0, the resistance change ΔR and the relative resistance change in the direction of the current flow Δ'_1 are given by

$$\Delta R = R - R_0, \quad \Delta'_1 = \frac{R - R_0}{R_0} = \frac{\rho'_1 - \rho_0}{\rho_0}. \tag{4.52}$$

The value for Δ'_1 relates to the stress as described by equation (4.33),

$$\Delta'_1 = \pi'_{11}\sigma'_x + \pi'_{12}\sigma'_y + \pi'_{13}\sigma'_z + \pi'_{14}\tau'_{yz} + \pi'_{15}\tau'_{xz} + \pi'_{16}\tau'_{xy}, \tag{4.53}$$

where we have used cartesian notation for the stress components according to Table 2.2.

Neglecting all Z'-components,

$$\Delta_1' = \pi_{11}'\sigma_x' + \pi_{12}'\sigma_y' + \pi_{16}'\tau_{xy}', \tag{4.54}$$

which provides the equation for the relative change of resistance for given values of longitudinal stress σ_x', transversal stress σ_y' and shear stress τ_{xy}'. The values for the transformed piezoresistance coefficients are calculated after rotation of the piezoresistive coefficients.

4.6.1 Example: Accelerometer with Cantilever and Piezoresistive Sensing

An accelerometer based on an inertial mass supported by a thin arm is shown in Figure 4.6. Both the inertial mass and the arm are made of p-type monocrystalline silicon of surface orientation (100). MEMS accelerometers are widely used in mobile phones, computer games and fitness apps such as pedometers, and are sensitive to a range of accelerations from 2g to 16g. Assuming that the cantilever has length $L = 1000\ \mu m$, width $W = 50\ \mu m$ and thickness $t = 2\ \mu m$, calculate the required value of an inertial mass at the tip of the cantilever in order to have a sensitivity at the output of a Wheatstone bridge of full range at an acceleration of 16g.

Supposing that the unit vectors of the cartesian coordinate system (X', Y', Z') of the mass and arm have unit vectors [011], [01$\bar{1}$] and [100] respectively, and that the crystallographic unit vectors of the silicon (X, Y, Z) are [011], [010] and [100] respectively, the rotation matrix is the same as in equation (4.40). We can calculate the values of the rotated piezoresistive coefficients:

$$\pi_{11}' = \pi_{11} - 2\pi_0\left(l_1^2 m_1^2 + l_1^2 n_1^2 + m_1^2 n_1^2\right) = \pi_{11} - \frac{\pi_0}{2} = \frac{1}{2}(-\pi_{11} + \pi_{12} + \pi_{44}),$$

$$\pi_{12}' = \pi_{12} + \pi_0\left(l_1^2 l_2^2 + n_1^2 n_2^2 + m_1^2 m_2^2\right) = \pi_{12} + \frac{\pi_0}{2} = \frac{1}{2}(\pi_{11} + \pi_{12} - \pi_{44}),$$

$$\pi_{16}' = 0. \tag{4.55}$$

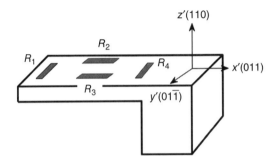

Figure 4.6 Piezoresistor accelerometer

In p-type silicon, we can approximate

$$\pi'_{11} \approx \frac{\pi_{44}}{2}, \quad \pi'_{12} \approx -\frac{\pi_{44}}{2}. \tag{4.56}$$

The cantilever will bend when subject to acceleration and hence the bending moment will be that calculated in equation (3.22), taking into account that in our geometry the force is downwards and hence negative:

$$M_b = -F(L - x). \tag{4.57}$$

Taking into account equations (3.4) and (3.14), the uniaxial stress along the cantilever length is

$$\sigma_x = -M_b \frac{y}{I_z}. \tag{4.58}$$

In this geometry the stress is tensile in the top surface of the cantilever and compressive in the bottom surface. The maximum stress occurs at both surfaces, $y = \pm t/2$, and at $x = 0$, thus

$$\sigma_x = FL \frac{t}{2I_z}. \tag{4.59}$$

With moment of inertia

$$I_z = \frac{Wt^3}{12}, \tag{4.60}$$

we obtain

$$\sigma_x = \frac{6FL}{Wt^2}. \tag{4.61}$$

In the geometry shown in Figure 4.6, the stress σ_x acts longitudinally on resistances R_2 and R_3 and transversally on resistances R_1 and R_4:

$$\left.\frac{\Delta R}{R}\right|_{2,3} = \Delta'_1 = \pi'_{11}\sigma'_x = \frac{\pi_{44}}{2}\sigma'_x,$$

$$\left.\frac{\Delta R}{R}\right|_{1,4} = \Delta'_2 = \pi'_{12}\sigma'_x = -\frac{\pi_{44}}{2}\sigma'_x. \tag{4.62}$$

The dimensions of the geometry in Figure 4.6 are not drawn to scale, and we assume that the four resistors are located very close to $x = 0$ and thus subject to the maximum stress. This result is very interesting as a differential measurement system takes advantage of it. We take the four resistors and arrange them in a Wheatstone bridge as shown in Figure 4.7. The output of the bridge is

$$V_0 = V_{cc} \frac{R_2 + \Delta R_2}{R_1 + \Delta R_1 + R_2 + \Delta R_2} - V_{cc} \frac{R_4 + \Delta R_4}{R_3 + \Delta R_3 + R_4 + \Delta R_4}, \tag{4.63}$$

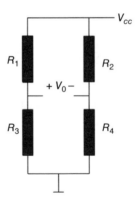

Figure 4.7   Example 4.6.1

which we can also write as

$$V_0 = V_{cc}\frac{R_2\left(1 + \frac{\Delta R_2}{R_2}\right)}{R_1\left(1 + \frac{\Delta R_1}{R_1}\right) + R_2\left(1 + \frac{\Delta R_2}{R_2}\right)} - V_{cc}\frac{R_4\left(1 + \frac{\Delta R_4}{R_4}\right)}{R_4\left(1 + \frac{\Delta R_4}{R_4}\right) + R_3\left(1 + \frac{\Delta R_3}{R_3}\right)}. \tag{4.64}$$

If the nominal values of the four resistors are equal, $R_1 = R_2 = R_3 = R_4$, substituting equations (4.62),

$$V_0 = \frac{V_{cc}}{2}\pi_{44}\sigma_x'. \tag{4.65}$$

Taking into account equation (4.61) and substituting the force $F = ma$, where m is the inertial mass and a is the acceleration,

$$V_0 = \frac{V_{cc}}{2}\pi_{44}\frac{6L}{Wt^2}ma. \tag{4.66}$$

As the goal of our design is to calculate the required value of the mass, we rearrange this to give

$$m = \frac{V_0}{a}\frac{2Wt^2}{6V_{cc}\pi_{44}L}, \tag{4.67}$$

where V_0/a is the sensitivity of the device. If we want full output for $a = 16g$, then $V_0 = V_{cc}$, and we obtain

$$m = 3.07 \times 10^{-7} \text{ kg}. \tag{4.68}$$

If we plan to fabricate the inertial mass with single-crystal silicon, the density of which is 2329 kg/m^3, the calculated value for the mass could be implemented with a silicon cube of side 509.1 μm.

4.7 Pressure Sensing with Rectangular Membranes

Rectangular membranes are used to sense the pressure using piezoresistive sensing. Proper design can be performed by analysing the deflection and stresses developing in the membrane for a given value of the differential pressure. The plate equation in two dimensions can be solved numerically for different boundary conditions and loads, as in [18]. The important thing to know is where the maximum values of the stress occur, for a given load, in order to place the piezoresistive sensors at those points.

Let us consider a square membrane as shown in Figure 4.8, in which the dimensions are normalized to the side length. The coordinates of the corners of the shaded quarter of membrane are also shown. Due to symmetry, only values of the stress are calculated inside the shaded area.

The results shown in [18] are adapted in Figure 4.9. The magnitude plotted is the normalized x-axis stress σ_{norm}, defined as

$$\sigma_{norm} = \sigma_x \frac{t^2}{PL^2}. \tag{4.69}$$

As can be seen, along the centre line of the membrane ($y/L = 0$), the stress parallel to the x-axis, σ_x, is normal to the edge of the membrane, and at $x/L = 0.5$ it has a maximum value σ_n:

$$\sigma_n = \sigma_x \left(\frac{x}{L} = 0.5, \frac{y}{L} = 0.5 \right) = 0.294 \frac{L^2}{t^2} P. \tag{4.70}$$

Similarly, the value of the x-axis stress at the edge $x/L = 0$ and $y/L = 0.5$ is a stress parallel to the membrane edge σ_p, and at this point is approximately (from Figure 4.9),

$$\sigma_p = \sigma_x \left(\frac{x}{L} = 0, \frac{y}{L} = 0.5 \right) = 0.115 \frac{L^2}{t^2} P. \tag{4.71}$$

Due to the membrane symmetry, equations (4.70) and (4.98) provide the values of the normal and parallel stresses at the centre of the sides of the membrane. At the edges, those values are

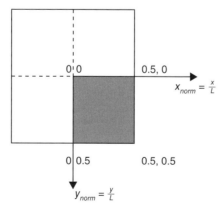

Figure 4.8 Pressure sensing with square membrane

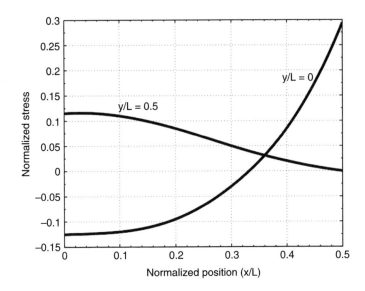

Figure 4.9 Plots of the normalized stress of equation (4.69) as a function of the x-position inside one quarter of the membrane for two values of the normalized y-position corresponding to the centre $y/L = 0$ and the membrane edge $y/L = 0.5$. Adapted from [18]

maximum. This information is very important for the design of pressure sensors as it indicates that the best choice for placing stress sensors is precisely in the centre of the edges of the membrane.

A typical example of a pressure sensor based on a square membrane clamped at the edges includes four piezoresistors in the centres of the four edges of the membrane (Figure 4.10).

As shown in Example 4.6.1, for p-type silicon (100) the longitudinal direction is the direction X' [011] and the transversal direction is Y' [01$\bar{1}$]. It can be assumed that

$$\pi_l = \pi'_{11} \approx \frac{\pi_{44}}{2}, \quad \pi_t = \pi'_{12} \approx -\frac{\pi_{44}}{2}. \tag{4.72}$$

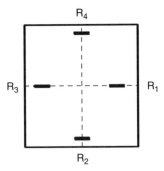

Figure 4.10 Four piezoresistors located at the edges of a square membrane

Table 4.2   Summary of stresses in Figure 4.10

	Longitudinal stress	Transversal stress
R_1 and R_3	normal to edge σ_n	parallel to edge σ_p
R_2 and R_4	parallel to edge σ_p	normal to edge σ_n

Looking at the membrane plot in Figure 4.10, two of the piezoresistors are parallel to the edge and the other two normal to the edge. In resistors R_1 and R_3, the longitudinal stress σ_n is normal to the edge, and the transversal stress σ_p is parallel to the edge. In contrast, in resistors R_2 and R_4 the longitudinal stress is parallel to the edge and the transversal stress is the normal to the edge. This is summarized in Table 4.2.

The relative change in value of the four resistors is

$$\frac{\Delta R_1}{R_1} = \frac{\Delta R_3}{R_3} = \Delta_1' = \pi_l' \sigma_n + \pi_t' \sigma_p = \frac{\pi_{44}}{2}(\sigma_n - \sigma_p), \tag{4.73}$$

$$\frac{\Delta R_2}{R_2} = \frac{\Delta R_4}{R_4} = \Delta_1' = \pi_l' \sigma_p + \pi_t' \sigma_n = \frac{\pi_{44}}{2}(\sigma_p - \sigma_n). \tag{4.74}$$

The four resistors are placed in a Wheatstone bridge as shown in Figure 4.11. A development of the circuit equations similar to that in Section 4.6.1 leads to the expression for the output voltage V_0:

$$V_0 = \frac{V_{cc}}{2}\pi_{44}(\sigma_p - \sigma_n). \tag{4.75}$$

From the definitions in equations (4.73) and (4.74),

$$V_0 = -0.179\frac{V_{cc}}{2}\frac{\pi_{44}L^2}{t^2}P. \tag{4.76}$$

For a pressure of 1000 Pa, membrane side length $L = 500\,\mu m$, membrane thickness $1\,\mu m$, $V_{cc} = 5$ V and $\pi_{44} = 138.1 \times 10^{-11}$ Pa, the value of the output voltage is $V_0 = -0.15$ V.

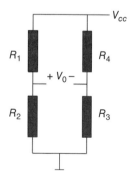

Figure 4.11   Wheatstone bridge

4.7.1 Example: Single-resistor Pressure Sensor

One of the most successful pressure sensors made by Motorola is the Xducer, a single piezoresistance located at the edge of a thin square membrane with the longitudinal orientation at an angle of 45° to the membrane edge [19, p. 243]. It is a good example of piezoresistive coefficient transformation. In fact there are three cartesian coordinate systems rotated with respect to each other around the Z-axis, as shown in Figure 4.12. The there coordinate systems are: (1) the membrane reference cartesian coordinate axes (XYZ), being the edges of the membrane along the X- and Y-axes; (2) the coordinate axes of the piezoresistance itself ($X'Y'Z'$), being the longitudinal dimension along the X'-axis, and the transversal along the Y'-axis; and (3) the crystallographic coordinate system of the silicon ($X''Y''Z''$). All three systems have the same Z-axis ($Z = Z' = Z''$). The piezoresistor is biased by a voltage V and the output is read as the voltage V_{out} generated across the piezoresistor transversal direction. We assume that the bias current is flowing only in the X'-direction and that the output voltage is measured with a high input impedance voltmeter. We can write Ohm's law for the resistor from equation (4.46), neglecting Z-components and denoting electric field components and current densities with primes to be consistent with the coordinate system of the resistor ($X'Y'Z'$):

$$\begin{pmatrix} \mathbf{E}'_x \\ \mathbf{E}'_y \end{pmatrix} = \begin{pmatrix} \rho'_1 & \rho'_6 \\ \rho'_6 & \rho'_2 \end{pmatrix} \begin{pmatrix} J'_x \\ 0 \end{pmatrix}. \tag{4.77}$$

Expanding,

$$\mathbf{E}'_x = \rho'_1 J'_x, \tag{4.78}$$

$$\mathbf{E}'_y = \rho'_6 J'_x. \tag{4.79}$$

The longitudinal and transversal components of the electric field are related to the bias and output voltages as follows:

$$\mathbf{E}'_x = \frac{V}{L}, \quad \mathbf{E}'_y = \frac{V_{out}}{W}, \tag{4.80}$$

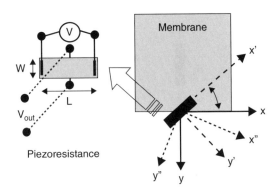

Figure 4.12 Sensor layout

and

$$V_{\text{out}} = V \frac{\rho'_6}{\rho'_1} \frac{W}{L}. \tag{4.81}$$

We see that the output voltage is proportional to the resistor aspect ratio, W/L, to the bias voltage and to the ratio of components 6 and 1 of the resistivity tensor. What we need is to bring into this equation the stresses through the piezoresistive coefficients. This can be done by recalling the definitions of the incremental resistance changes,

$$\Delta'_1 = \frac{\rho'_1 - \rho_0}{\rho_0}, \quad \Delta'_6 = \frac{\rho'_6}{\rho_0}. \tag{4.82}$$

Substituting into equation (4.81),

$$\frac{\rho'_6}{\rho'_1} = \frac{\Delta'_6}{1 + \Delta'_1} \approx \Delta'_6 \tag{4.83}$$

and

$$V_{\text{out}} = V \Delta'_6 \frac{W}{L}. \tag{4.84}$$

From the relationship between the stress components and the resistance changes, it follows that

$$(\Delta') = (\Pi') \begin{pmatrix} \sigma'_x \\ \sigma'_y \\ 0 \\ 0 \\ 0 \\ \tau'_{xy} \end{pmatrix}, \tag{4.85}$$

where we only consider the components of the stress in the plane. As we only need to find Δ'_6, we can expand equation (4.85) to obtain

$$\Delta'_6 = \pi'_{61} \sigma'_x + \pi'_{62} \sigma'_x + \pi'_{66} \tau'_{xy} \tag{4.86}$$

and

$$V_{\text{out}} = V \frac{W}{L} (\pi'_{61} \sigma'_x + \pi'_{62} \sigma'_x + \pi'_{66} \tau'_{xy}). \tag{4.87}$$

Let us now transform the components of the stress. We will follow the procedure described in Section 2.9. The rotation matrix should transform the components of the stress in the coordinate axes of the membrane (XYZ) into the components in the axes of the piezoresistance ($X'Y'Z'$):

$$(R) = \begin{pmatrix} \cos(X'X) & \cos(X'Y) & \cos(X'Z) \\ \cos(Y'X) & \cos(Y'Y) & \cos(Y'Z) \\ \cos(Z'X) & \cos(Z'Y) & \cos(Z'Z) \end{pmatrix} \begin{pmatrix} \cos\theta & -\sin\theta & 0 \\ \sin\theta & \cos\theta & 0 \\ 0 & 0 & 1 \end{pmatrix}, \qquad (4.88)$$

where $\cos(XX')$ denotes the cosine of the angle between the X-axis and the X'-axis. For the stress transformation, we assume that we only have a longitudinal stress component along the X-axis, σ_x. This transforms into the three stress components in the coordinate system of the piezoresistance (($\sigma'_x, \sigma'_y, \tau'_{xy}$), assuming all Z components are negligible) as follows:

$$(T') = (R)(T)(R)^T \qquad (4.89)$$

$$= \begin{pmatrix} T'_1 & T'_6 & T'_5 \\ T'_6 & T'_2 & T'_4 \\ T'_5 & T'_4 & T'_3 \end{pmatrix} \begin{pmatrix} \sigma'_x & \tau'_{xy} & 0 \\ \tau'_{xy} & \sigma'_y & 0 \\ 0 & 0 & 0 \end{pmatrix} \qquad (4.90)$$

$$= \begin{pmatrix} \cos\theta & -\sin\theta & 0 \\ \sin\theta & \cos\theta & 0 \\ 0 & 0 & 1 \end{pmatrix} \begin{pmatrix} \sigma_x & 0 & 0 \\ 0 & 0 & 0 \\ 0 & 0 & 0 \end{pmatrix} \begin{pmatrix} \cos\theta & \sin\theta & 0 \\ -\sin\theta & \cos\theta & 0 \\ 0 & 0 & 1 \end{pmatrix}. \qquad (4.91)$$

Expanding,

$$\sigma'_x = \cos^2\theta\,\sigma_x, \quad \sigma'_y = \sin^2\theta\,\sigma_x, \quad \tau'_{xy} = -\sin\theta\cos\theta\,\sigma_x. \qquad (4.92)$$

The last step is to relate the piezoresistance coefficients in the coordinate axis of the silicon crystal (X'', Y'', Z'') to the coordinate system of the piezoresistance (X', Y', Z'), as shown in Figure 4.13.

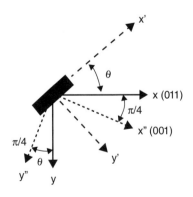

Figure 4.13 Rotation of (X', Y', Z') to (X'', Y'', Z'')

The new rotation matrix (R') is easily calculated from the angles shown in Figure 4.13 and using the same equation (4.88) where the axes $(X''Y''Z'')$ replace the axes (XYZ):

$$R' = \begin{pmatrix} l_1 & m_1 & n_1 \\ l_2 & m_2 & n_2 \\ l_3 & m_3 & n_3 \end{pmatrix} = \begin{pmatrix} \cos\left(\theta + \frac{\pi}{4}\right) & \cos\left(\frac{3\pi}{4} + \theta\right) & \cos\frac{\pi}{2} \\ \cos\left(\frac{\pi}{4} - \theta\right) & \cos\left(\theta + \frac{\pi}{4}\right) & \cos\frac{\pi}{2} \\ \cos\frac{\pi}{2} & \cos\frac{\pi}{2} & \cos 0 \end{pmatrix} \quad (4.93)$$

The transformation between the piezoresistance coefficients is

$$\pi'_{61} = \pi_0 \left(l_1^3 l_2 + m_1^3 m_2 + n_1^3 n_2 \right),$$

$$\pi'_{62} = \pi_0 \left(l_1 l_2^3 + m_1 m_2^3 + n_1 n_2^3 \right),$$

$$\pi'_{66} = \pi_{44} - 2\pi_0 \left(l_1^2 l_2^2 + m_1^2 m_2^2 \right). \quad (4.94)$$

Substituting the values of the l's, m's and n's, we obtain

$$\pi'_{61} = \pi_0(\cos^2\theta - \sin^2\theta)(\sin\theta - \cos\theta),$$

$$\pi'_{62} = -\pi_0(\sin\theta + \cos\theta)(\sin\theta - \cos\theta),$$

$$\pi'_{66} = \pi_{44} - \pi_0(\cos^2\theta - \sin^2\theta). \quad (4.95)$$

Substituting into equation (4.87) and after some algebraic manipulation we obtain

$$V_0 = V\frac{W}{2L}\pi_{44}\sigma_x \sin 2\theta. \quad (4.96)$$

The sensitivity is maximized when $\theta = \pi/4$.

4.7.2   Example: Pressure Sensors Comparison

Compare the output voltage of two pressure sensors based on square membrane, one with four resistors as in Figure 4.10 and the other with a single resistor as in Figure 4.12, and with angle $\theta = 45°$.

Let us divide equation (4.96) by equation (4.76) to compare the output voltages of both sensors,

$$\frac{V_{\text{single}}}{V_{\text{four}}} = \frac{V(W/2L)\pi_{44}\sigma_x \sin 2\theta}{-0.179(V_{cc}/2)\pi_{44}(L^2/t^2)P}. \quad (4.97)$$

Comparing for equal values of the bias voltage $V = V_{cc}$ and taking into account that the longitudinal stress σ_x in the single-resistor device is parallel to the edge stress, we can write

$$\sigma_p = 0.115\frac{L^2}{t^2}P, \quad (4.98)$$

and then

$$\frac{V_{\text{single}}}{V_{\text{four}}} = -\frac{0.115}{0.179}\frac{W}{L}\sin 2\theta. \tag{4.99}$$

If we select the optimum angle for the single-resistor device $\theta = 45°$,

$$\frac{V_{\text{single}}}{V_{\text{four}}} = -\frac{0.115}{0.179}\frac{W}{L}. \tag{4.100}$$

Note that W and L are the dimensions of the single piezoresistor, and normally resistors are made such that $W/L < 1$. The output voltage of the single-resistor device will thus be smaller that the output of the four-resistor device.

4.8 Piezoelectricity

As briefly summarized in Section 1.6, piezoelectricity is concerned with the interaction of the electric domain with the mechanical domain in a material. If we have an unstressed medium, the relationship between the field strength $\mathbf{E}$ and the electric displacement is $D = \epsilon\mathbf{E}$, where ϵ is the permittivity. If we have a stressed medium with no applied electric field, the relationship between the strain S and the stress T is simply $[S] = [s][T]$, as in Chapter 2, where $[s]$ is the compliance matrix. In piezoelectric materials the two domains interact and, to a good approximation, it can be assumed that the interaction is linear. The direct piezoelectric effect is described by

$$D = dT + \epsilon\mathbf{E} \tag{4.101}$$

where d is the piezoelectric coupling tensor. The inverse piezoelectric effect is described by

$$S = sT + d^T\mathbf{E}, \tag{4.102}$$

where d^T is the transpose of the piezoelectric coupling tensor. In an anisotropic medium equation (4.101)expands to give

$$\begin{pmatrix} D_1 \\ D_2 \\ D_3 \end{pmatrix} = \begin{pmatrix} d_{11} & d_{12} & d_{13} & d_{14} & d_{15} & d_{16} \\ d_{21} & d_{22} & d_{23} & d_{24} & d_{25} & d_{26} \\ d_{31} & d_{32} & d_{33} & d_{34} & d_{35} & d_{36} \end{pmatrix} \begin{pmatrix} T_1 \\ T_2 \\ T_3 \\ T_3 \\ T_5 \\ T_6 \end{pmatrix} + \begin{pmatrix} \epsilon_{11} & \epsilon_{21} & \epsilon_{31} \\ \epsilon_{12} & \epsilon_{22} & \epsilon_{32} \\ \epsilon_{13} & \epsilon_{23} & \epsilon_{33} \end{pmatrix} \begin{pmatrix} \mathbf{E}_1 \\ \mathbf{E}_2 \\ \mathbf{E}_3 \end{pmatrix}.$$

$$\tag{4.103}$$

The permittivity tensor is always symmetric. If the coordinate axes are the principal axes of the material, then the permittivity tensor is diagonal. In a uniaxial material such as quartz, two

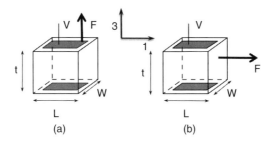

Figure 4.14 Modes of operation: (a) axial; (b) transversal

of the diagonal elements are equal and the third is different. The inverse piezoelectric equation for anisotropic materials can be written as

$$
\begin{pmatrix} S_1 \\ S_2 \\ S_3 \\ S_4 \\ S_5 \\ S_6 \end{pmatrix} = \begin{pmatrix} s_{11} & s_{12} & s_{13} & s_{14} & s_{15} & s_{16} \\ s_{21} & s_{22} & s_{23} & s_{24} & s_{25} & s_{26} \\ s_{31} & s_{32} & s_{33} & s_{34} & s_{35} & s_{36} \\ s_{41} & s_{42} & s_{43} & s_{44} & s_{45} & s_{46} \\ s_{51} & s_{52} & s_{53} & s_{54} & s_{55} & s_{56} \\ s_{61} & s_{62} & s_{63} & s_{64} & s_{65} & s_{66} \end{pmatrix} \begin{pmatrix} T_1 \\ T_2 \\ T_3 \\ T_4 \\ T_5 \\ T_6 \end{pmatrix} + \begin{pmatrix} d_{11} & d_{21} & d_{31} \\ d_{12} & d_{22} & d_{32} \\ d_{13} & d_{23} & d_{33} \\ d_{14} & d_{24} & d_{34} \\ d_{15} & d_{25} & d_{35} \\ d_{16} & d_{26} & d_{36} \end{pmatrix} \begin{pmatrix} \mathbf{E}_1 \\ \mathbf{E}_2 \\ \mathbf{E}_3 \end{pmatrix}.
$$

$$(4.104)$$

Piezoelectric actuators can operate in two modes: axial and transversal. In axial mode (Figure 4.14(a)) the actuation has the same direction of the applied electric field, whereas in the transversal mode (Figure 4.14(b)) the actuation takes place orthogonally to the electric field direction.

According to conventional nomenclature, in axial mode, the electric field has the direction of axis 3 and the actuation is also in this same axis 3. Since only the component $\mathbf{E}_3$ of the electric field is different from zero and only the component T_3 of the stress tensor is different from zero, equation (4.104) simplifies to

$$S_3 = s_{33}T_3 + d_{33}\mathbf{E}_3. \qquad (4.105)$$

These piezoelectric actuators can operate without any restraint (so $T_3 = 0$), and then the maximum value of the strain is

$$S_3\big|_{\max} = d_{33}\mathbf{E}_3. \qquad (4.106)$$

The actuation stroke is

$$\Delta t = S_3 t, \qquad (4.107)$$

and the maximum stroke is

$$\Delta t\big|_{\max} = d_{33}\mathbf{E}_3 t. \qquad (4.108)$$

If the actuator works against a restraint, then it develops a force F that in equilibrium is equal to the restraint and in the opposite direction. The maximum value of the force is when $S_3 = 0$ and

$$0 = -\frac{s_{33}F_{max}}{A} + d_{33}\mathbf{E}_3,$$

and then

$$F_{max} = \frac{d_{33}\mathbf{E}_3 A}{s_{33}} = \frac{d_{33}\mathbf{E}_3 WL}{s_{33}}. \qquad (4.109)$$

As the electric field is $\mathbf{E}_{33} = V/t$,

$$F_{max} = \frac{d_{33}WL}{s_{33}t}V. \qquad (4.110)$$

Sometimes an effective stiffness k_{eff} at constant electric field is defined as

$$k_{eff} = \frac{F_{max}}{\Delta t|_{max}} = \frac{WL}{s_{33}t}. \qquad (4.111)$$

In the transversal operation mode, the electric field has the direction of the axis 3 whereas the displacement takes place in the direction of axis 1. Equation (4.104) simplifies to

$$S_1 = s_{11}T_1 + d_{31}\mathbf{E}_3. \qquad (4.112)$$

Similarly to the derivation above for the axial mode, here the maximum stroke of the actuator is found from equation (4.102) when $T_1 = 0$,

$$\Delta L|_{max} = d_{31}\mathbf{E}_3 L. \qquad (4.113)$$

The maximum force when $S_1 = 0$ and $T_1 = -F_{max}/Wt$ is

$$F_{max} = \frac{d_{31}W}{s_{11}}V, \qquad (4.114)$$

and the effective stiffness is given by

$$k_{eff} = \frac{Wt}{s_{11}L}. \qquad (4.115)$$

4.8.1   Relevant Data for Some Piezoelectric Materials

The main parameters involved in the stroke and maximum force equations of piezoelectric actuators are the piezoelectric coupling coefficients d_{31} and d_{33} and the compliance matrix elements s_{11} and s_{33}. Typical data for these four parameters for some piezoelectric materials are summarized in Table 4.3.

Table 4.3   Summary of piezoelectric parameter values [27, 28]

Material	$d_{31} \times 10^{-12}$ C/N	$d_{33} \times 10^{-12}$ C/N	$s_{11} \times 10^{-12}$ Pa^{-1}	$s_{33} \times 10^{-12}$ Pa^{-1}
ZnO	−5.43	11.67	7.86	6.94
PZT-4	−123	289	12.3	15.5
AlN	−1.9	4.9	0.0031	0.0027

4.8.2   Example: Piezoelectric Generator

In energy-harvesting applications the force exerted in a cantilever by an external acceleration is converted into an output electric voltage. In Figure 4.15 a cantilever made of a metallic film is covered on both sides by piezoelectric material film. Calculate the voltage generated when the applied force deflects the tip of the cantilever by 2 μm. The cantilever length is L = 400 μm, the width is W = 100 μm, the piezoelectric material is AlN and has a thickness of t_2 = 1 μm and the thickness of the metallic strip is t_1 = 0.2 μm. The piezoelectric coefficient is d_{31} = −1.9 pC/N. Data have been taken from [29].

We know from Chapter 3 that the deflection of a cantilever clamped at one end subject to a point force at the tip is given by

$$v = \frac{F}{EI}\left(L\frac{x^2}{2} - \frac{x^3}{6}\right). \tag{4.116}$$

The maximum value of the deflection for a given force occurs at the tip, $x = L$, and the value is $v(x = L) = v_{max}$. Equation (4.116) can be written as

$$v = \frac{3v_{max}x^2}{L^3}\left(\frac{L}{2} - \frac{x}{6}\right), \tag{4.117}$$

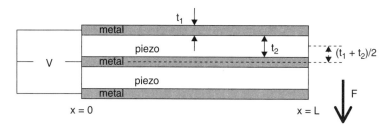

Figure 4.15   Example 4.8.2. Cross-section of a bimorph made of two piezo layers and three metal layers. The left-hand end is fixed and the force is assumed to be applied at the right-hand end. The neutral axis is the long dashed line

with

$$v_{\text{max}} = \frac{FL^3}{3EI}. \tag{4.118}$$

We also know that the strain S inside the cantilever is a linear function of the distance y to the neutral plane,

$$S = -\frac{M_b}{EI}y. \tag{4.119}$$

The bending moment is

$$M_b = F(L - x), \tag{4.120}$$

hence

$$S = -\frac{F}{EI}(L - x)y. \tag{4.121}$$

We can substitute F/EI in terms of v_{max} from equation (4.118),

$$S = -\frac{3v_{\text{max}}}{L^3}(L - x)y. \tag{4.122}$$

Due to symmetry, the neutral axis lies in the middle and the stress is linearly related to distance to the neutral line. The stress is positive above the neutral line and negative below. The electric field generated has opposite signs in both piezoelectric layers and hence the electrical connection shown in Figure 4.15 is a parallel connection. In order to evaluate the average electric field generated, we calculate the average strain at the mid-point inside the piezoelectric film, at $y = (t_1 + t_2)/2$ and then average it over the entire length of the cantilever:

$$S_{\text{avg}} = -\frac{t_1 + t_2}{2L} \int_0^L \frac{3v_{\text{max}}}{L^3}(L - x)dx = -\frac{t_1 + t_2}{4}\frac{3v_{\text{max}}}{L^2}. \tag{4.123}$$

The average strain is related to the electric field that is generated by the force as

$$S_{\text{avg}} = d_{31}\mathbf{E}, \tag{4.124}$$

where $\mathbf{E}$ is the electric field. In this example, as $\mathbf{E} = V/t_2$,

$$v_{\text{max}} = \frac{4}{3}\frac{L^2 d_{31}}{(t_1 + t_2)t_2}V. \tag{4.125}$$

For a tip deflection of $v_{\text{max}} = 2\ \mu\text{m}$, the voltage generated is $V = 5.68$ V.

Problems

4.1 Demonstrate the analytical expression for the rotated piezoresistive coefficient π'_{11}.

4.2 The National Highway Traffic Safety Administration (NHTSA) recommends a deceleration of $60g$ to deploy airbags, although the driver's weight and other factors are also taken into consideration by the electronic trigger system. In this problem we would like to evaluate the performance of an accelerometer made of a cantilever supporting an inertial mass at the tip and a piezoresistor located at the cantilever end closest to the support.

   Assuming that the inertial mass is a silicon volume of 2.5×10^{-12} m^3, calculate the force when the acceleration is $60g$. Considering this as a point force applied to the tip of the cantilever itself, calculate the deflection of the tip as a function of the length L and the value for $L = 1000\,\mu$m. The cantilever is made of silicon, and has width $W = 50\,\mu$m and thickness $t = 2\,\mu$m. Assume that the edge of the cantilever is clamped. Young's modulus for silicon is $E = 164$ GPa.

4.3 For the same data as in Problem 4.2, calculate the distribution of stress along the cantilever length as a function of the position y inside the cantilever above the neutral surface. Calculate the value of the maximum stress and identify the location where this occurs.

4.4 For the same data as in Problem 4.2, if a piezoresistor is placed where the maximum stress occurs, calculate the value of the relative change in the resistance when the device is subject to $100g$ acceleration. Plot the values of the deflection, stress and $\Delta R/R$ as a function of the cantilever length L.

4.5 Calculate the output of a Wheatstone bridge such as the one shown in Example 4.6.1 of piezoresistances made of polysilicon film, located as shown in Figure 4.16 on top of a membrane clamped in the periphery.

4.6 We place four resistors in a Wheatstone bridge as in Example 4.6.1. Write the equation giving the output voltage as a function of the relative increment of the resistors after applying a pressure difference of P. Calculate the value of the output voltage when $P = 1000$ Pa, $\ell = 500\,\mu$m, $h = 1\,\mu$m, and $\pi_{44} = 138 \times 10^{-11}$.

4.7 We have a pressure sensor based on a circular membrane of radius a and we would like to compare the sensitivity provided by a piezoresistor oriented radially at the edge

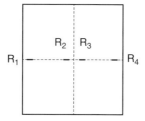

Figure 4.16   Problem 4.5

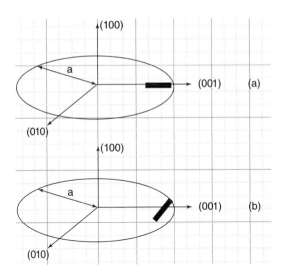

Figure 4.17 Problem 4.7

of the membrane (Figure 4.17(a)) with a piezoresistor oriented tangentially also at the edge of the membrane (Figure 4.17(b)).

The membrane is made out of n-type silicon (100) having longitudinal piezoresistance coefficient $\pi_l = -102, 2 \times 10^{-11}$ Pa^{-1}, and transversal piezoresistance coefficient $\pi_t = 53.4 \times 10^{-11}$ Pa^{-1}. These values correspond to the crystallographic coordinates, $v = 0.33$, $a = 500\,\mu$m and $h = 1\,\mu$m.

Calculate the relative change of resistance when the differential pressure P applied is 10^4 Pa for the two cases. What position of the piezoresistance gives better sensitivity?

4.8 We have a clamped circular membrane of thickness $h = 2\,\mu$m and radius $a = 200\,\mu$m. The Poisson ratio is $v = 0.27$. For a pressure difference of $P = 1000$ Pa, calculate the maximum of the radial stress and the maximum of the absolute value of the tangential stress. Let us put a piezoresistor along the radial direction at the point of maximum radial stress. We know that this direction is (011). Using the results of the piezoresistive coefficients for these rotated axes from Example 4.5.1, calculate the value of the relative change of the piezoresistor ($\Delta R/R$).

4.9 We have a bimorph made of a stack of two materials, one of them piezoelectric. If we bias the piezoelectric material with a voltage V, the electric field created (assumed to be in direction 3) is V/d, where d is the thickness of the piezoelectric material, which is a fraction of the total thickness of the bimorph. Calculate the equation giving the stress in the direction of axis 1. Calculate the bending moment corresponding to this stress and solve the beam equation to find the deflection as a function of the position and the deflection at the tip.

5

Electrostatic Driving and Sensing

This chapter covers the fundamentals of electrostatic sensing and actuation. The first part of the chapter describes a generic parallel-plate actuator and how the electrostatic force between plates can be calculated when a voltage is applied. Details of the movement of the plates allows the definition of the pull-in and pull-out conditions, and a simple dynamic model is described allowing the calculation of switching speed and power consumption from the source. Lateral actuator geometry is also described, leading to the comb actuator structure. Differential capacitive sensing is introduced and its application to acceleration measurement described. The chapter concludes with a description of the torsional actuator.

5.1 Energy and Co-energy

A large number of applications arise in microelectromechanical devices rooted in the actuation and sensing properties of separate electrodes subject to an electric field or to mechanical forces. Many of the fundamental design equations as well as the operation of switches, capacitive sensors and tunable capacitors can be introduced through the study of a simple system of two parallel plates, as shown in Figure 5.1. The product of the charge Q and voltage V in this system has units of work. A state variable space can be drawn relating these two magnitudes, as shown in Figure 5.2 for an element having an arbitrary relationship between charge and voltage (e.g. a nonlinear capacitor).

The differential of the product of the two magnitudes is

$$d(VQ) = VdQ + QdV, \tag{5.1}$$

and after integration

$$VQ = \int VdQ + \int QdV = U + U^*, \tag{5.2}$$

where U is the energy and U^* is the co-energy [30].

Understanding MEMS: Principles and Applications, First Edition. Luis Castañer.
© 2016 John Wiley & Sons, Ltd. Published 2016 by John Wiley & Sons, Ltd.
Companion Website: www.wiley.com/go/castaner/understandingmems

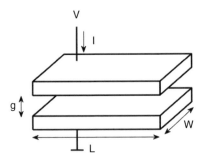

Figure 5.1   Parallel-plate actuator

Let us consider a parallel-plate system as shown in Figure 5.3. We suppose that the plate initially located at the origin can move, whereas the plate located at $x = g_0$ cannot move. When electrical work is done, either by applying a voltage V or by supplying an electrical charge Q, compensating mechanical work also has to be done to preserve equilibrium.

In a differential of time dt, the electrical work is the product of the power and time and, as the power is the product of the voltage V and current I, the electrical work is $V \times I \times dt$. We assume that a force F is applied to hold the plate steady in equilibrium. The mechanical work is the product of the force F and the velocity v, and the velocity is dx/dt. Taking into account that $I \times dt = dQ$,

$$dU = d(VQ) = VIdt + Fvdt = VdQ + Fdx. \tag{5.3}$$

In a conservative system, the integral of dU along a closed loop has to be zero,

$$\oint dU = 0, \quad \text{hence} \quad \oint (VdQ + Fdx) = 0. \tag{5.4}$$

Equation (5.4) requires the integrand to be a complete differential, hence

$$dU = \left(\frac{\partial U}{\partial Q}\right)_x dQ + \left(\frac{\partial U}{\partial x}\right)_Q dx. \tag{5.5}$$

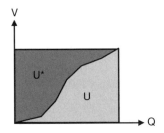

Figure 5.2   State variable space

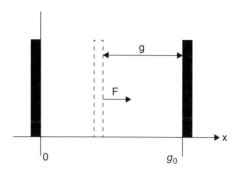

Figure 5.3   Parallel-plate geometry

Comparing equations (5.3) and (5.5),

$$V = \left(\frac{\partial U}{\partial Q}\right)_x, \quad F = \left(\frac{\partial U}{\partial x}\right)_Q. \tag{5.6}$$

Similarly, for the co-energy,

$$dU^* = d(VQ) - dU = QdV - Fdx, \quad dU^* = \left(\frac{\partial U^*}{\partial V}\right)_x dV + \left(\frac{\partial U^*}{\partial x}\right)_V dx, \tag{5.7}$$

and

$$Q = \left(\frac{\partial U^*}{\partial V}\right)_x, \quad F = -\left(\frac{\partial U^*}{\partial x}\right)_V. \tag{5.8}$$

As can be seen, there are two definitions of force: as partial derivative of the energy at constant charge (equation 5.6); and as partial derivative of the co-energy at constant voltage with negative sign (equation 5.8). This helps in computing the force under different driving scenarios where charge or voltage can be assumed constant in the transition from one state to another.

Let us now explore how the operating point of the system moves from one initial state to another final state. The state space is depicted in Figure 5.4. Let us consider a final state

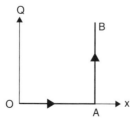

Figure 5.4   Integration of dU in the (Q, x) space

characterized by a charge stored Q and by a change in the position x of the plate (point B). The initial state is considered to be at the origin of the (Q, x) space (point O).

In a conservative system the result of the integration of the dU between two points (the initial point O in Figure 5.4, and the final point B in Figure 5.4) is independent of the path of integration. We select the path $O \Rightarrow A \Rightarrow B$, therefore the charge is $Q = 0$ between O and A and mechanical work has to be applied to split apart the two plates from $x = 0$ to $x = x$. In the path from A to B, the distance between the plates is constant and the charge increases from $Q = 0$ to $Q = Q$. Integrating dU from O to B gives

$$U = \int_O^B dU = \int_O^A VdQ + \int_A^B VdQ + \int_O^A Fdx + \int_A^B Fdx. \tag{5.9}$$

As between O and A the initial and final values of Q are zero, and as between A and B the initial and final values of x are the same, it follows that only the second integral is different from zero,

$$U = \int_A^B VdQ = \int_0^Q VdQ. \tag{5.10}$$

Voltage and charge are related by the capacitance $C(x)$,

$$V = \frac{Q}{C(x)}, \tag{5.11}$$

and as the integration from A to B is done for constant x, $C(x)$ is constant and can be taken outside the integral,

$$U = \int_0^Q \frac{QdQ}{C(x)} = \frac{1}{C(x)} \int_0^Q QdQ, \tag{5.12}$$

yielding

$$U = \frac{Q^2}{2C(x)}, \quad \text{with } C(x) = \frac{\epsilon_0 A}{g_0 - x}, \tag{5.13}$$

where A is the area of the plate. Similarly, the integration of the differential of the co-energy dU^* can be done in a similar state space (V, x),

$$dU^* = QdV - Fdx \quad \text{and} \quad U^* = \int_0^V QdV = \frac{1}{2}C(x)V^2. \tag{5.14}$$

From the two definitions of the force, the correct one depends on the driving method: at constant voltage, equation (5.6) should be used; but at constant charge, equation (5.8) should be used instead.

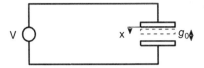

Figure 5.5 Voltage drive

5.2 Voltage Drive

The most common way to drive an electrostatic actuator based on a parallel-plate geometry is to apply a voltage source V, as shown in Figure 5.5. Generally the charging time of the capacitor is much shorter than the mechanical time constant of the movable microstructure, and the capacitor becomes charged before any movement of the plates has taken place. Let us suppose that one of the plates (upper plate) is movable (e.g. it is supported by elastic beams), while the other is fixed.

The voltage applied to the capacitor exerts a force on the movable plate that has to be compensated by an external force to ensure equilibrium. The external force can be calculated using equation (5.8),

$$F = - \left(\frac{\partial U^*}{\partial x} \right)_V , \tag{5.15}$$

and taking into account equation (5.14),

$$F_E = -F = \frac{V^2}{2} \frac{\partial C}{\partial x} = \frac{V^2}{2} \frac{\epsilon_0 A}{(g_0 - x)^2} = \frac{\epsilon_0 A V^2}{2g^2}. \tag{5.16}$$

Since $Q = CV$ and $\epsilon_0 A = C_0 g_0$, with $C_0 = \epsilon_0 A / g_0$, the electrostatic force can also be written as

$$F_E = \frac{Q^2}{2\epsilon_0 A} = \frac{Q^2}{2C_0 g_0}. \tag{5.17}$$

As can be seen, the electrostatic force F_E has the sign of the axis, indicating that the force is attractive between electrodes. Moreover, the force is proportional to the square of the voltage applied. As a result the force is always attractive, irrespective of the sign of the applied voltage. It can also be seen that the force is inversely proportional to $(g_0 - x)^2$, so it increases as the gap decreases. This situation is in fact a positive feedback that leads to the collapse of the upper plate. This is called the pull-in instability.

5.3 Pull-in Voltage

Due to the nonlinear nature of the equations describing the equilibrium of the plate electrode, the question arises as to whether all equilibrium points are stable [31]. In order to analyse the equilibrium conditions and their stability, we consider the geometry depicted in Figure 5.6, where the upper plate of a parallel-plate actuator is held by a spring and we assume that the

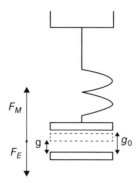

Figure 5.6 Parallel-plate geometry

bottom plate is fixed. When a voltage V is applied between the two plates an electrostatic force F_E develops downwards and the upper plate is held in equilibrium by means of the upward spring restoring force F_M. In equilibrium the two forces must be equal. The initial gap is g_0 and the remaining gap, at the moment where the equilibrium happens, is g. The electrostatic and mechanical forces are

$$F_E = \frac{V^2}{2} \frac{\epsilon_0 A}{g^2} \quad \text{and} \quad F_M = k(g_0 - g), \tag{5.18}$$

where k is the stiffness of the spring or flexure. The equilibrium condition is

$$F_E = F_M, \quad \text{so that} \quad \frac{V^2}{2} \frac{\epsilon_0 A}{g^2} = k(g_0 - g). \tag{5.19}$$

The second equation that must be written concerns the stability of the equilibrium points defined by the solution of equation (5.19). This condition can be written as

$$\frac{d(F_E - F_M)}{dg} > 0. \tag{5.20}$$

Equation (5.20) means that an instantaneous increase of g must be compensated by an increase in the downward force:

$$\frac{d}{dg} \left(\frac{V^2}{2} \frac{\epsilon_0 A}{g^2} - k(g_0 - g) \right) > 0 \tag{5.21}$$

or

$$-\frac{V^2 \epsilon_0 A}{g^3} + k > 0. \tag{5.22}$$

Rearranging equation (5.19),

$$\epsilon_0 A V^2 = 2g^2 k(g_0 - g), \tag{5.23}$$

and substituting into equation (5.22),

$$g > \frac{2}{3}g_0. \tag{5.24}$$

Equation (5.24) means that only values of g greater than $\frac{2}{3}g_0$ provide a stable position for the upper plate, restricting the stable range. Of course if $g < \frac{2}{3}g_0$ then the upper plate will collapse on top of the bottom electrode as no points are stable in this range. The equations above provide the necessary means to find the voltage value required to reach pull-in. Substituting $g = \frac{2}{3}g_0$ in equation (5.22), it follows that the voltage, called the pull-in voltage V_{PI}, is given by

$$V_{\text{PI}} = \sqrt{\frac{8kg_0^3}{27\epsilon_0 A}}. \tag{5.25}$$

5.3.1 Example: Forces in a Parallel-plate Actuator

Calculate the values of the mechanical restoring force and electrostatic force in a parallel-plate actuator as a function of the remaining gap for applied voltages of 1 V, 2 V, 3 V and 4 V. The plates have an area of $500 \times 500\ \mu m$, the initial gap is $g_0 = 4\ \mu m$ and the stiffness constant of the spring is $k = 1.6\ N/m$. Calculate the value of the pull-in voltage, calculate the equilibrium points for $V = 3\ V$, and identify the stable equilibrium point.

Figure 5.7 shows a plot of the electrostatic and mechanical forces as a function of the remaining gap for the four voltage values. As can be seen, for 4 V there is no intersection between the mechanical and electrostatic forces, indicating that there are no equilibrium points. However, the plots for 1, 2 and 3 V intersect with the mechanical force plot at two points each. Considering the requirements for an equilibrium point to be stable given in equation (5.24), just one of the points of intersection will be stable. The pull-in voltage is given by

$$V_{\text{PI}} = \sqrt{\frac{8 \times 1.6 \times (4 \times 10^{-6})^3}{27 \times 8.85 \times 10^{-12} \times (500 \times 10^{-6})^2}} = 3.7\,\text{V}. \tag{5.26}$$

To find the equilibrium points we have to solve equation (5.19), which can be written in polynomial form,

$$2kg^3 - 2kg_0 g^2 + \epsilon_0 A V^2 = 0, \tag{5.27}$$

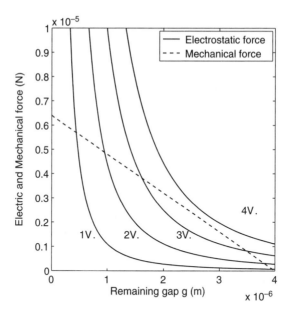

Figure 5.7   Parallel-plate geometry

and substituting the data values,

$$g^3 - 4.09 \times 10^{-5} g^2 + 6.21 \times 10^{-18} = 0. \tag{5.28}$$

Equation (5.28) can be solved using MATLAB code:

```
c1=1;
c2=-4.09e-5;
c3=0;
c4= 6.21e-18;
p=[c1 c2 c3 c4];
r=roots(p);
```

and the solution is the vector r

```
r =
    1.0e-05
    0.3809
    0.0953
   -0.0762
```

One of the three roots of the polynomial is negative, so the two intersection points are the positive solutions $g_1 = 3.089$ μm and $g_2 = 0.953$ μm. Of these two values g_1 is the only one greater than $\frac{2}{3} g_0 = 2.66$ μm, and hence the only stable point.

5.4 Electrostatic Pressure

In a parallel-plate actuator the attractive electrostatic force between the plates results in a pressure

$$P_E = \frac{F_E}{A} = \frac{\epsilon_0 A V^2}{2g^2 A} = \frac{\epsilon_0 V^2}{2g^2} = \frac{\epsilon_0}{2}E^2. \tag{5.29}$$

This equation helps set a limit for the operation of devices if air breakdown has to be avoided. The breakdown electric field is approximately 10^6 V/m and so the maximum electrostatic pressure is in the region of 45 kPa.

5.5 Contact Resistance in Parallel-plate Switches

Pull-in-based switches can also be used in DC or low-frequency applications to replace solid-state switches. The contact resistance is then of prime importance. It is known that the contact resistance between two metals is given by [32]

$$R_{ON} = \rho \left(\frac{6E}{32 F_C r_C} \right)^{1/3}, \tag{5.30}$$

where ρ is the resistivity of the metal, F_C is the contact force, E is Young's modulus and r_C is the radius of curvature of a spherical elastic contact. As can be seen, the value of the ON resistance becomes large for small contact forces. If the total available force is distributed over a number N of parallel contacts, the total resistance becomes

$$R_{ON} = \frac{\rho}{2(4F_{tot} r_C / 3E)^{1/3}} \frac{1}{N^{2/3}}. \tag{5.31}$$

5.6 Hold-down Voltage

In switching applications of parallel-plate actuators, one way to avoid plate short-circuits is to deposit a thin dielectric film of thickness t on top of the bottom plate, as shown in Figure 5.8.

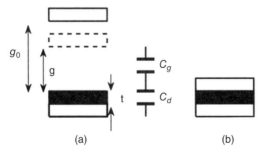

Figure 5.8   (a) Up position and (b) down position

Once the upper plate has collapsed on top of the dielectric film, if we start to decrease the voltage applied, the plate will remain down until the restoring mechanical force is larger than the downward electrostatic force. We can write this condition by noting that the capacitance C of the two series capacitors C_d and C_g is given by

$$\frac{1}{C} = \frac{1}{C_g} + \frac{1}{C_d} = \frac{g_0}{\epsilon_0 A} + \frac{t}{\epsilon_0 \epsilon_r A}. \tag{5.32}$$

Hence

$$C = \frac{A\epsilon_0}{g + t/\epsilon_r}. \tag{5.33}$$

The actuator can thus be considered as having a capacitance of the same area and an effective gap equal to $g + t/\epsilon_r$. Then the electrostatic force can be written as

$$F_E = \frac{\epsilon_0 A V^2}{2(g + t/\epsilon r)^2}. \tag{5.34}$$

The hold-down condition can be written as

$$k(g_0 - g) \geq \frac{\epsilon_0 A V^2}{2(g + t/\epsilon_r)^2}. \tag{5.35}$$

Taking equality in (5.35), the hold-down V_{HD} voltage can be calculated:

$$V_{HD} = \left(g + \frac{t}{\epsilon_r}\right)\sqrt{\frac{2k(g_0 - g)}{\epsilon_0 A}}\Bigg|_{g=0} = \frac{t}{\epsilon_r}\sqrt{\frac{2kg_0}{\epsilon_0 A}}. \tag{5.36}$$

5.6.1   Example: Calculation of Hold-down Voltage

Calculate the value of the hold-down voltage for a parallel-plate actuator with plate area $500 \times 500\ \mu m$, initial gap $g_0 = 4\ \mu m$, spring stiffness constant $k = 1.6\ N/m$, dielectric thickness $t = 200\ nm$ and dielectric relative permittivity 3.9.

Using equation (5.36),

$$V_{HD} = \frac{200 \times 10^{-9}}{3.9}\sqrt{\frac{2 \times 1.6 \times 4 \times 10^{-6}}{8.85 \times 10^{-12} \times 25 \times 10^{-8}}} = 0.12\ \text{V}. \tag{5.37}$$

5.7   Dynamic Response of Pull-in-based Actuators

The dynamic response can be modelled by a spring–mass–dashpot system as shown in Figure 5.9 and described by the equations [33]

$$m\frac{d^2x}{dt^2} + b\frac{dx}{dt} + kx = \frac{Q^2}{2C_0 g_0}, \qquad \frac{dQ}{dt} = \frac{V_s - V}{R}. \tag{5.38}$$

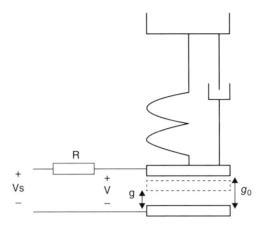

Figure 5.9   Spring–mass–dashpot model

The first equation is the electromechanical equation and the second is the circuit equation. We consider here an internal resistor R of the voltage source. The two equations are coupled because

$$Q = CV = \frac{\epsilon_0 A}{g} V \quad \text{and} \quad g = g_0 - x. \tag{5.39}$$

The dynamic response can be calculated by solving the differential equations above using a numerical algorithm. Taking into account that the state variables of the problem are the displacement x, the velocity $v = dx/dt$ and the charge Q, the formulation for numerical solution is as follows:

$$\frac{dx}{dt} = v,$$

$$\frac{dv}{dt} = -2\omega_0 \zeta v - \omega_0^2 x + \frac{Q^2}{2 C_0 g_0 m},$$

$$\frac{dQ}{dt} = \frac{V_s}{R} - \frac{Q(g_0 - x)}{R C_0 g_0}, \tag{5.40}$$

where $\omega_0 = \sqrt{k/m}$ and $\zeta = b/2\sqrt{km}$.

5.7.1   Example: Switching Transient

Find the switching time of a parallel-plate actuator having the following parameter values:
$b = 1.7 \times 10^{-3}$, $m = 15.4 \times 10^{-6}$, $k = 4.6$, $g_0 = 8 \times 10^{-6}$, $V_S = 10\,V$, $R = 1\,\Omega$, $A = 1.2 \times 10^{-6}$
and $\epsilon_0 = 8.85 \times 10^{-12}$.

To solve the problem we can write MATLAB code as follows:

```
function dy=dynamics(t,y)
b=1.7e-3
m=15.4e-6
k=4.6
go=8e-6
vs=10
R=1
A=1.2e-6
epsilon0=8.85e-12
Co=epsilon0*A/go
```

We use here an ordinary differential equation solver (ode15s) that calls the dynamics function:

```
function dy=dynamics(t,y)
dy=zeros(3,1)
dy(1)=y(2);
dy(2) =-(b/m)*y(2)-(k/m)*y(1)+y(3)^2 /(2*Co*go*m);
dy(3)=(vs/R)-y(3)*(go-y(1))/(R*Co*go)
```

where y is the vector of the three state variables and dy is the vector of first derivatives of the three state variables. The result is shown in Figure 5.10, where the switching time can be estimated to be 5×10^{-3} s.

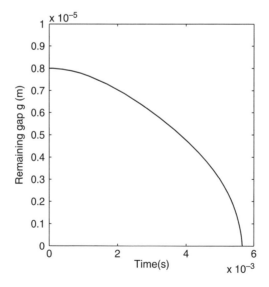

Figure 5.10   Remaining gap g as a function of time in Example 5.7.1

5.8   Charge Drive

A charge drive can be implemented in such a way that a given amount of charge is supplied to the parallel-plate actuator sufficiently fast that the movable plate does not have time to move until all charge is in the capacitor. This drive is for Q constant, therefore the value of the force F can be calculated using equations (5.8) and (5.19),

$$F = \left(\frac{\partial U}{\partial x}\right)_Q \quad \text{and} \quad C(x) = \frac{\epsilon_0 A}{g_0 - x}. \tag{5.41}$$

As

$$U = \frac{Q^2}{2C} \tag{5.42}$$

and Q is constant, we obtain

$$F_E = -F = \frac{Q^2}{2\epsilon_0 A}, \tag{5.43}$$

which is a result entirely similar to the voltage drive case. Again the force is attractive but only depends on the amount of charge squared, so it is independent of the sign of the charge.

5.9   Extending the Stable Range

The parallel-plate actuators described in the sections above suffer from the pull-in instability that reduces the range of travel of the movable electrode. Although this is not a problem for digital applications where we want the upper plate to collapse, it may be a problem for analogue applications such as variable capacitors as the stable range of gap values is restricted from g_0 to $\frac{2}{3}g_0$. One of the possible solutions to this problem is to implement a charge drive of the actuator instead of a voltage drive [34]. The equilibrium condition is

$$F_E = F_M \quad \text{and} \quad \frac{Q^2}{2\epsilon_0 A} = k(g_0 - g), \tag{5.44}$$

and the stability condition is

$$\frac{d(F_E - F_M)}{dg} > 0 \quad \text{and} \quad \frac{d}{dg}\left(\frac{Q^2}{2\epsilon_0 A} - k(g_0 - g)\right) > 0. \tag{5.45}$$

Obviously, and because Q is constant, the left-hand side of the second equation in (5.45) is always k which is always positive. This means that, theoretically, the pull-in instability can be avoided by implementing a charge drive (see Problems 5.3 and 5.4).

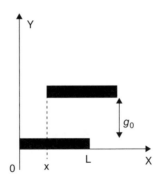

Figure 5.11   Lateral actuator geometry

5.10   Lateral Electrostatic Force

In many devices the desired movement is not one plate against the other but sliding parallel to the other (Figure 5.11). The force between the plates can be written

$$F_E = \frac{V^2}{2} \frac{\partial C}{\partial x}. \tag{5.46}$$

As the movement is restricted to the lateral displacement, the capacitance is

$$C = \frac{\epsilon_0 (L - x) W}{2 g_0}, \tag{5.47}$$

and hence

$$F_E = -\frac{\epsilon_0 V^2 W}{2 g_0}. \tag{5.48}$$

The resulting force is to the left. The general rule is that the movement tends to make the superposition area between plates the largest possible. Furthermore, the force is independent of the relative position of the two plates and only depends on the gap and on the voltage applied squared. Lateral force is the basic principle of comb actuators.

5.11   Comb Actuators

The basic structure of a comb actuator is shown in Figure 5.12. Two fingers (1 and 2) of the comb actuator are shown and can move in the up or down direction. The upper electrode has three fingers and cannot move. The gap between the movable fingers and the fixed electrodes is g_0. In the absence of an applied voltage, the superimposed area between electrodes is $y_0 t$. The total capacitance is

$$C_0 = 4 \frac{\epsilon_0 y_0 t}{g_0} \tag{5.49}$$

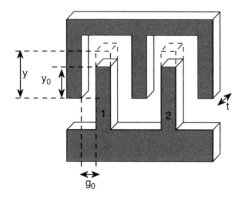

Figure 5.12 Comb drive geometry

as there are four identical capacitors in parallel. The total capacitance when the electrodes are located in the dashed position in Figure 5.12 is

$$C = 4\frac{\epsilon_0 y t}{g_0}. \tag{5.50}$$

The electrostatic force is given by

$$F_E = -\left(\frac{\partial U^*}{\partial g}\right)_V = -\frac{\partial}{\partial g}\left(\frac{CV^2}{2}\right)_V = -\frac{2\epsilon_0 V^2 t}{g_0}. \tag{5.51}$$

If the spring or flexure holding the movable fingers has a stiffness k, the mechanical force is

$$F_M = k(y - y_0). \tag{5.52}$$

In equilibrium at an applied voltage V,

$$\frac{2\epsilon_0 V^2 t}{g_0} = k(y - y_0) \quad \text{so} \quad y = y_0 + \frac{2\epsilon_0 V^2 t}{kg_0}. \tag{5.53}$$

In practical realizations N fingers are used and the capacitance is given by

$$C = 2N\frac{\epsilon_0 y t}{g_0}. \tag{5.54}$$

The equilibrium displacement y for a given voltage V is

$$y = y_0 + \frac{N\epsilon_0 V^2 t}{kg_0}. \tag{5.55}$$

Large displacements can be achieved using a large number of fingers.

5.12   Capacitive Accelerometer

When a microelectromechanical system is subject to an acceleration, the simplest model is a lumped model similar to equation (5.38). The independent term is the external force given by $F = ma_{ext}$, where a_{ext} is the external acceleration:

$$m\frac{d^2x}{dt^2} + b\frac{dx}{dt} + kx = ma_{ext}. \tag{5.56}$$

For a sinusoidal permanent regime, the Laplace transform of the transfer function is

$$H(s) = \frac{X(s)}{A_{ext}(s)} = \frac{m}{ms^2 + bs + k}, \tag{5.57}$$

where $A_{ext}(s)$ is the Laplace transform of the time variant acceleration a_{ext}. Using the same definitions for ω_0 and for ζ as in Section 5.7 and for $s = j\omega$,

$$H(j\omega) = \frac{1}{-\omega^2 + j2\zeta\omega_0\omega + \omega_0^2}, \quad |H(j\omega)| = \frac{1}{\sqrt{\left(\omega_0^2 - \omega^2\right)^2 + (2\zeta\omega_0\omega)^2}}. \tag{5.58}$$

The displacement amplitude is

$$\widehat{x} = \frac{\widehat{a_{ext}}}{\sqrt{\left(\omega_0^2 - \omega^2\right)^2 + (2\zeta\omega_0\omega)^2}}, \tag{5.59}$$

where $\widehat{x}$ is the amplitude of the displacement and $\widehat{a_{ext}}$ is the amplitude of the acceleration. Normally accelerometers work at frequencies much lower than the resonance frequency $\omega \ll \omega_0$ and equation (5.59) simplifies to

$$\widehat{x} = \frac{\widehat{a_{ext}}}{\omega_0^2}. \tag{5.60}$$

As can be seen, the measurement of acceleration can be made by means of the measurement of displacement as the two magnitudes are proportional at low frequencies. However, the proportionality constant is $1/\omega_0^2$ and a very sensitive accelerometer would involve a low value of the resonant frequency, and hence a slow one. So there is a trade-off between sensitivity and speed.

5.13   Differential Capacitive Sensing

Displacement can be measured capacitively, because if the displacement increases or reduces the distance between two parallel electrodes, the capacitance will also change accordingly. Commercial accelerometers [35] are based on this principle. Of course displacement can also be measured using parallel displacement of plates at constant distance. In this case the capacitance change is proportional to the relative displacement.

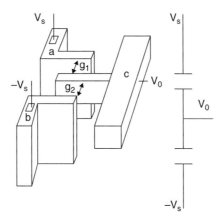

Figure 5.13 Differential capacitor

In this section we analyse the case shown in Figure 5.13, where a movable electrode (c) connected to an inertial mass is placed between two fixed electrodes (a and b). Movement of the middle electrode is allowed only towards the fixed electrodes and the three electrodes remain parallel to each other. The gaps g_1 and g_2 change when there is a movement of the inertial mass. The electrical equivalent circuit (also shown in Figure 5.13) involves two capacitors connected in the middle electrode. When it moves, the capacitances C_{ab} and C_{bc}, between electrodes a and b and between electrodes b and c respectively, can be written

$$C_{ab} = \frac{\epsilon_0 A}{g_1}, \quad C_{bc} = \frac{\epsilon_0 A}{g_2}, \tag{5.61}$$

where A is the area of the plates. If two voltages of the same amplitude and opposite phase are connected to the outer electrodes, such as $V_s \sin \omega t$ and $V_s \cos \omega t$, the circuit equation in the Laplace transformed domain is

$$V_0 = -V_s + 2V_s \frac{\frac{1}{sC_2}}{\frac{1}{sC_2} + \frac{1}{sC_1}} = V_s \left(-1 + \frac{2C_1}{C_1 + C_2} \right). \tag{5.62}$$

Substituting equation (5.61),

$$V_0 = V_s \frac{g_2 - g_1}{g_1 + g_2}. \tag{5.63}$$

Taking into account that $g_1 = g_0 + \delta$ and $g_2 = g_0 - \delta$, where δ is the displacement,

$$V_0 = V_s \frac{\delta}{g_0}. \tag{5.64}$$

The transduction from displacement into amplitude of the sinusoidal signal is linear.

5.14   Torsional Actuator

In applications it is sometimes necessary to have a tilted plane that can move on a hinge when a voltage is applied. A simplified layout of this torsional actuator is shown in Figure 5.14 [36].
The capacitance of a differential section dx around x can be written as

$$dC = \frac{\epsilon_0 A}{y} = \frac{\epsilon_0 W dx}{y}, \tag{5.65}$$

where W is the width of the plate. From Figure 5.14,

$$\tan \theta = (y - d)/x, \tag{5.66}$$

and then

$$dC = \frac{\epsilon_0 W dx}{d + x \tan \theta}. \tag{5.67}$$

If a voltage V is applied between the electrodes (the tilted plate and the bottom shaded area), the electrostatic force is given by

$$dF = -\left.\frac{\partial(dU^*)}{\partial y}\right|_V, \tag{5.68}$$

and as

$$dU^* = \frac{V^2 dC}{2},$$

we obtain

$$dF = \frac{\epsilon_0 W V^2}{2(d + x \tan \theta)^2} dx. \tag{5.69}$$

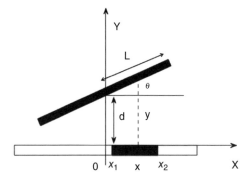

Figure 5.14   Torsional actuator

The differential force creates a moment

$$M = \int_{x_1}^{x_2} dFr dx,$$ (5.70)

where $r = x/\cos\theta$ is the distance between the point of application of the dF and the centre of rotation. Then

$$M = \int_{x_1}^{x_2} \frac{V^2\epsilon_0 Wx}{2(d + x\tan\theta)^2 \cos\theta} dx.$$ (5.71)

Changing variable $z = d + x\tan\theta$, with $z_1 = d + x_1\tan\theta$ and $z_2 = d + x_2\tan\theta$,

$$M = \frac{\epsilon_0 WV^2}{2\cos\theta} \int_{z_1}^{z_2} \frac{z - d}{(z\tan\theta)^2} dx = \frac{\epsilon_0 WV^2}{2\sin\theta\tan\theta} \left(\ln\left(1 + \frac{b\tan\theta}{d}\right) + \frac{d}{d + \tan\theta} - 1\right),$$

(5.72)

where we have approximated $x_1 = 0$ and $b = x_2 - x_1$.

To find the equilibrium point, the moment calculated must be equal to the torsion moment $k_\theta\theta$, where k_θ is the torsion stiffness. These actuators also suffer from pull-in if driven by a voltage source. The value of the pull-in voltage has been calculated as [37]

$$V_{PI} \approx \sqrt{0.827\frac{k_\theta d^2}{\epsilon_0 WL^3}},$$ (5.73)

where $L = d/\sin\theta_{max}$ and θ_{max} is the maximum torsion angle occurring when the tip of the torsional plate of length L is grounded.

Problems

5.1 For the same data as in Example 5.3.1, plot the values of g as a function of the applied voltage.

5.2 Calculate the total energy consumed from the source during the switching of an actuator of the same parameter values as in Example 5.3.1.

5.3 One way to extend the travel range of pull-in-based actuators is to add an external capacitor C_{ext} in series with the actuator as shown in Figure 5.15. Show that the condition required to get the full stable range is that $C_0/C_{ext} > 2$, where $C_0 = \epsilon_0 A/g_0$. Calculate the equation giving the voltage V that should be applied to get full gap stability.

5.4 For the same circuit as in Problem 5.3, we have a MEMS switch of area $200 \times 200\ \mu m^2$, $g_0 = 1\ \mu m$, $k = 5.2$ N/m. (1) Calculate the required value of the external capacitance C_{ext}. (2) Calculate the value of the voltage required and compare it with the pull-in voltage of the MEMS switch.

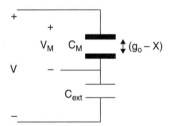

Figure 5.15 Problem 5.3

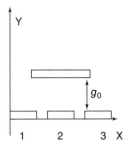

Figure 5.16 Problem 5.6

5.5 A MEMS parallel-plate actuator voltage driven is driven beyond pull-in. To avoid short-circuiting the voltage source, a thin dielectric is deposited on top of one of the electrodes. (1) Calculate the value of the capacitance when the switch is open. (2) Calculate the capacitance when the switch is closed. (3) Calculate the tunable range C_{ON}/C_{OFF}. Take $g_0 = 2\ \mu m$, $t = 200\ nm$, $\epsilon_r = 3.9$ and $\epsilon_0 = 8.85 \times 10^{-12}$ F/m.

5.6 A linear electrostatic motor is based on a distribution of electrodes such as shown in Figure 5.16. (1) Indicate the sign of the movement when a voltage is applied between electrode 1 and the upper electrode. (2) Indicate the sign of the movement when the voltage is applied between electrode 3 and the upper electrode. (3) Calculate the strength of the force in both cases. Take $g_0 = 2\ \mu m$, $\epsilon_0 = 8.85 \times 10^{-12}$ F/m, $W = 20\ \mu m$ and $V = 10$ V.

5.7 A parallel-plate MEMS switch is used to implement a voltage DC/DC step-up converter.[1] The switch is first driven beyond pull-in, the upper plate collapses and the capacitance is charged. A short-circuit is avoided by depositing a thin dielectric film on top of the bottom plate. Then a mechanical pull force is applied to raise the plate to the up position. The diode shown in Figure 5.17 blocks the discharge of the capacitor in this phase. Calculate the amount of charge received by the capacitor in the charging phase. Calculate the value of the voltage across the capacitor after the pulling phase has ended. Take $g_0 = 2\ \mu m$, $t = 100\ nm$, $\epsilon_r = 3.9$, $\epsilon_0 = 8.85 \times 10^{-12}$ F/m and area $200 \times 200\ \mu m^2$.

---

[1] C.H. Haas, M. Kraft, Modelling and analysis of a MEMS approach to dc voltage step-up conversion, J. Micromech. and Microeng., 14, pp. 114–122, 2004.

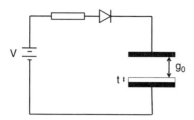

Figure 5.17 Problem 5.7

5.8 Solve the differential equation for the dynamic response of the parallel-plate actuator if a delta pulse of charge is applied at $t = 0$ with a value of $Q_s = 8 \times 10^{11}$ C, and find the time variation of the deflection for $t > 0$. The data values are: $b = 0.17$, $m = 15.4 \times 10^{-6}$, $k = 4.6$, $g_0 = 8 \times 10^{-6}$, $A = 1.2 \times 10^{-6}$, $\epsilon_0 = 8.85 \times 10^{-12}$ and $Q_S = 8 \times 10^{-11}$. Recall that $C_0 = \epsilon_0 A / g_0$.

5.9 We would like to use a torsional actuator, such as the one depicted in Figure 5.12, to measure angle by measuring capacitance. Calculate the maximum tilt angle we can have. Calculate the change in capacitance when the change in angle is from $\theta_{max}/2$ to $\theta_{max}/3$. Take $L = 15\,\mu m$, $b = 12\,\mu m$, $d = 5\,\mu m$.

5.10 In an electrostatic actuator of two parallel plates moving against each other, we want to know the condition required to have just one equilibrium point. The plates have a surface of $A = 200\,\mu m \times 200\,\mu m$, the air permittivity is 8.85×10^{-12} F/m, the initial gap is $g_0 = 2\,\mu m$ and $k = 1$ N/m. Write the condition required in order to have a single equilibrium point. Calculate the value of the gap g satisfying this requirement. Is this point stable? Calculate the voltage required to reach this equilibrium point.

5.11 We have two ways of implementing a movement to close a gap as shown in Figure 5.18, torsional actuation or a comb actuator, and we are interested in calculating the voltage required in the two cases and also the total energy drawn from the source. Take $y_0 = 1\,\mu m$, $t = 1\,\mu m$, $k = 1$ N/m, $\epsilon_0 = 8.85 \times 10^{-12}$ F/m, $g_0 = 1\,\mu m$, $d = 2\,\mu m$, $L = 5\,\mu m$, $W = 12\,\mu m$, $k_\theta = 8 \times 10^{-14}$ N/mrad. We use a circuit to drive the actuator, composed of a DC voltage source V_{CC} having the same value as the voltage required to close the gap and a series resistor $R = 1\,k\Omega$.

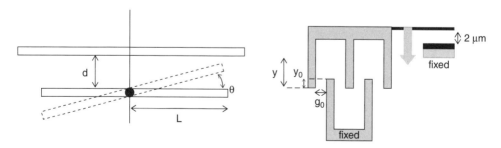

Figure 5.18 Problem 5.11

5.12 We have a comb actuator and we want to calculate the number of fingers required to have a displacement of 20 μm when the applied voltage is 20V. The data values are $y_0 = 200$ μm, $t = 2$ μm, $g_0 = 1$ μm and $k = 0.01$ N/m.

5.13 We have an accelerometer based on the measurement of displacement described in Section 5.12. If we set a target value of $\omega_0 = 2\pi \times 10^3$ rad/s, simulate and plot the transfer function $|H(j\omega)|$ as a function of the frequency.

5.14 With the same data as in Problem 5.13, find the transient response of the accelerometer to a pulse of 10 ms^{-2} and plot it as a function of time.

5.15 For the same value of $\omega_0 = 2\pi \times 10^3$ rad/s as in Problem 5.13, find the critical damping condition and plot the transient response to an acceleration pulse of 10 ms^{-2} as a function of time.

6

Resonators

Resonance is used in MEMS technology for sensing purposes as the resonance frequency is sensitive to various magnitudes such as the mass of the vibrating microstructure or the stress. This is useful for gravimetric sensors detecting the change in mass due to adsorption or for resonant accelerometers or gyroscopes. The main purpose of this chapter is to review the basic concepts of resonance using a lumped parameter model for definitions and to study some of the properties by using a PSpice model. This is applied in order to find the equivalent circuit of an electrostatically driven resonator, and PSpice is used to illustrate the series and resonance frequencies. The Raleigh–Ritz method for finding the resonance frequency using energy methods is illustrated to find the resonance of cantilevers and its application to gravimetric chemical sensors. The principles of gyroscopes based on the Coriolis acceleration are then described and an implementation using a tuning fork illustrated.

6.1 Free Vibration: Lumped-element Model

Resonance can be defined using a simplifying lumped model assumption for the the inertia and elastic components of the force. The simplest case is that of an undamped system. If we define x as the deflection of the point mass, we have

$$m\frac{d^2x}{dt^2} + kx = 0, \quad \text{with } \omega_0 = \sqrt{\frac{k}{m}}, \tag{6.1}$$

which we can rearrange as

$$\frac{d^2x}{dt^2} + \omega_0^2 x = 0. \tag{6.2}$$

The solution of this second-order differential equation is

$$x = A\sin(\omega_0 t + \phi), \tag{6.3}$$

Understanding MEMS: Principles and Applications, First Edition. Luis Castañer.
© 2016 John Wiley & Sons, Ltd. Published 2016 by John Wiley & Sons, Ltd.
Companion Website: www.wiley.com/go/castaner/understandingmems

where ω_0 is the resonant frequency. If the initial conditions are set to zero the solution is trivially zero. If an initial value of the deflection $x(t = 0) = x_0$ and of its first time derivative $(dx/dt)_{t=0}$ are assumed, the solution is

$$A = \sqrt{x_0^2 + \frac{(\frac{dx}{dt})_{t=0}^2}{\omega_0^2}}, \quad \phi = \tan^{-1}\left(\frac{x_0\omega_0}{(\frac{dx}{dt})_{t=0}}\right). \tag{6.4}$$

The resonant angular frequency ω_0 can also be found using energy methods by considering that, in this conservative system, the maximum of the potential energy must be equal to the maximum of the kinetic energy. The potential energy U_P and the kinetic energy U_K are given by

$$U_P = \frac{1}{2}kx^2, \quad U_K = \frac{1}{2}m\left(\frac{dx}{dt}\right)^2. \tag{6.5}$$

The maximum of the potential energy occurs when $x = A$ and the maximum of the kinetic energy occurs when the velocity is maximum $(dx/dt = A\omega_0)$:

$$U_{P_{max}} = \frac{1}{2}kA^2 \quad U_{K_{max}} = \frac{1}{2}mA^2\omega_0^2. \tag{6.6}$$

It follows that

$$\omega_0 = \sqrt{\frac{k}{m}}, \tag{6.7}$$

which is the expression for the resonance frequency.

6.2 Damped Vibration

Vibration also exists when there is damping. Damping modifies the differential equation by adding a term proportional to the velocity,

$$m\frac{d^2x}{dt^2} + b\frac{dx}{dt} + kx = 0, \quad \text{with } \omega_0 = \sqrt{\frac{k}{m}} \text{ and } \zeta = \frac{b}{2\sqrt{km}}, \tag{6.8}$$

which we can rearrange as

$$\frac{d^2x}{dt^2} + 2\zeta\omega_0\frac{dx}{dt} + \omega_0^2x = 0. \tag{6.9}$$

The roots of the characteristic equation

$$r^2 + 2\zeta\omega_0 r + \omega_0 = 0 \tag{6.10}$$

are

$$r_1 = -\zeta\omega_0 + \omega_0\sqrt{\zeta^2 - 1} \quad \text{and} \quad r_2 = -\zeta\omega_0 - \omega_0\sqrt{\zeta^2 - 1}. \tag{6.11}$$

The vibration is underdamped if $\zeta < 1$, and then the roots are complex conjugates,

$$r_1 = -\zeta\omega_0 + j\omega_0\sqrt{1 - \zeta^2} \quad \text{and} \quad r_2 = -\zeta\omega_0 - j\omega_0\sqrt{1 - \zeta^2}. \tag{6.12}$$

The underdamped solution is

$$x(t) = Ae^{r_1 t} + Be^{r_2 t} = e^{-\zeta\omega_0 t}(Ae^{j\omega_d} + Be^{-j\omega_d}), \tag{6.13}$$

with

$$\omega_d = \omega_0\sqrt{1 - \zeta^2}. \tag{6.14}$$

The solution shown is equivalent to $Ce^{-\zeta\omega_0 t}\sin(\omega_d t + \phi)$, thus the transient that develops is a damped sinusoidal function of time having the angular frequency ω_d, called the damping vibration frequency.

6.3 Forced Vibration

In inertial MEMS devices a time-dependent force is applied. A particular case of interest is when the applied force is a sinusoidal function of time. After a initial transient from the initial conditions, a sinusoidal permanent regime is reached. Let us suppose that a time-dependent force $F(t)$ is applied to the spring–mass system with damping described above,

$$m\frac{d^2x}{dt^2} + b\frac{dx}{dt} + kx = F(t). \tag{6.15}$$

Taking the Laplace transforms yields

$$m\left(s^2X(s) - sx(0) - \frac{dx}{dt}\Big|_{t=0}\right) + b(sX(s) - x(0)) + kX(s) = F(s). \tag{6.16}$$

If the initial conditions are assumed to be zero, $x(0) = 0$ and $dx/dt|_{t=0} = 0$, then

$$X(s)(ms^2 + bs + k) = F(s). \tag{6.17}$$

A transfer function between the output $X(s)$ and the input $F(s)$ can be defined as

$$H(s) = \frac{X(s)}{F(s)} = \frac{1}{ms^2 + bs + k}. \tag{6.18}$$

Using the definitions of ω_0 and ζ,

$$\omega_0 = \sqrt{\frac{k}{m}} \quad \text{and} \quad \zeta = \frac{b}{2\sqrt{km}}, \tag{6.19}$$

the transfer function becomes

$$H(s) = \frac{1/m}{s^2 + 2\zeta\omega_0 s + \omega_0^2}. \tag{6.20}$$

In the sinusoidal permanent regime, the applied force is $F(t) = F_0 \sin \omega t$ and the amplitude of the output $\hat{x}$ is equal to the product of the amplitude of the input function, F_0, and the modulus of $H(j\omega)$, $|H(j\omega)|$. Substituting $s = j\omega$ in equation (6.20),

$$H(j\omega) = \frac{1/m}{-\omega^2 + j2\zeta\omega_0\omega + \omega_0^2}, \quad |H(j\omega)| = \frac{1/m}{\sqrt{\left(\omega_0^2 - \omega^2\right)^2 + (2\zeta\omega_0\omega)^2}}. \tag{6.21}$$

The deflection is

$$x = \hat{x}\sin(\omega t + \phi), \tag{6.22}$$

where

$$\hat{x} = \frac{F_0/m}{\sqrt{\left(\omega_0^2 - \omega^2\right)^2 + (2\zeta\omega_0\omega)^2}} \quad \text{and} \quad \tan^{-1}\phi = -\frac{2\zeta\omega_0\omega}{\omega_0^2 - \omega^2}. \tag{6.23}$$

In equation (6.23) it can be seen that at low frequencies $\omega \ll \omega_0$,

$$\hat{x} \to F_0/k. \tag{6.24}$$

At low frequencies, the amplitude of the deflection is equal to the steady-state deflection caused by a force of magnitude F_0. However, for intermediate frequencies, the output amplitude has a maximum that can be found by setting

$$\frac{d\hat{x}}{d\omega} = 0,$$

which occurs at frequency

$$\omega_r = \omega_0\sqrt{1 - 2\zeta^2}. \tag{6.25}$$

This maximum requires $2\zeta^2 < 1$; when $2\zeta^2 = 1$ or, equivalently, $\zeta = \frac{1}{\sqrt{2}}$, the system is in 'critical damping'. The value of the amplitude at $\omega = \omega_r$ can be calculated from equations (6.23) and (6.25),

$$\hat{x}(\omega = \omega_r) = \frac{F_0/m}{\sqrt{\left(\omega_0^2 - \omega_0^2(1 - 2\zeta^2)\right)^2 + \left(2\zeta\omega_0^2\sqrt{1 - 2\zeta^2}\right)^2}}, \tag{6.26}$$

which simplifies to

$$\hat{x}(\omega = \omega_r) = \frac{F_0}{2k\zeta\sqrt{1 - \zeta^2}}. \tag{6.27}$$

Remember that this equation holds provided that $2\zeta^2 < 1$.

The frequency ω_r is called the resonant frequency of a forced vibration of a damped system. There are, then, three main frequencies involved: the free vibration of the undamped system, ω_0; the free vibration frequency of the damped system ω_d; and the frequency of the forced vibration at maximum amplitude, ω_r. It can be seen that they are close to each other provided that the value of the parameter ζ is small compared to unity. A quality factor Q is defined as the peak value of the relative amplitude of the output at the resonant frequency ω_r,[1]

$$Q = \frac{\hat{x}(\omega_r)}{F_0/k} = \frac{1}{2\zeta\sqrt{1 - \zeta^2}}. \tag{6.28}$$

The behaviour of equation (6.23) for practical resonators working at a frequency close to resonance indicates that the function has very low values except for frequencies very close to resonance (see example below). For this reason, equation (6.23) can be simplified by substituting $\omega \simeq \omega_0$ in all terms with the exception of the term $(\omega_0^2 - \omega^2)$, thus

$$\hat{x} = \frac{F_0/m}{\sqrt{\left(\omega_0^2 - \omega^2\right)^2 + \left(2\zeta\omega_0^2\right)^2}}. \tag{6.29}$$

Normalizing equation (6.29) to the maximum value,

$$\hat{x}_{norm} = \frac{2\omega_0^2\zeta\sqrt{1 - \zeta^2}}{\sqrt{\left(\omega_0^2 - \omega^2\right)^2 + \left(2\zeta\omega_0^2\right)^2}}. \tag{6.30}$$

Setting equation (6.30) equal to $1/\sqrt{2}$, the -3 dB angular frequencies ω_1 and ω_2 can be found:

$$\hat{x}_{norm} = \frac{2\omega_0^2\zeta\sqrt{1 - \zeta^2}}{\sqrt{\left(\omega_0^2 - \omega^2\right)^2 + \left(2\zeta\omega_0^2\right)^2}} = \frac{1}{\sqrt{2}}. \tag{6.31}$$

---

[1] Q is used here to dessignate the quality factor, not to be confused with the elcetrical charge Q used in previous chapters

Simplifying

$$\omega_0^2 - \omega^2 = (\omega_0 + \omega)(\omega_0 - \omega) \simeq 2\omega_0(\omega_0 - \omega) \tag{6.32}$$

yields

$$\hat{x}_{\text{norm}}^2 = \frac{4\omega_0^4 \zeta^2 (1 - \zeta^2)}{(2\omega_0(\omega_0 - \omega))^2 + (2\zeta\omega_0^2)^2} = \frac{1}{2}, \tag{6.33}$$

leading to

$$(\omega_0 - \omega)^2 = \zeta^2 \omega_0^2 (1 - 2\zeta^2) \tag{6.34}$$

and

$$\omega_0 - \omega = \pm\zeta\omega_0 \sqrt{1 - 2\zeta^2}. \tag{6.35}$$

Calculating the two solutions ω_1 and ω_2, the -3 dB bandwidth $\omega_1 - \omega_1$ is

$$\omega_2 - \omega_1 = 2\zeta\omega_0 \sqrt{1 - 2\zeta^2}. \tag{6.36}$$

An alternative definition of the quality factor Q is the ratio of the resonant frequency and the -3 dB bandwidth,

$$Q = \frac{\omega_r}{\omega_2 - \omega_1} = \frac{\omega_0 \sqrt{1 - 2\zeta^2}}{\omega_2 - \omega_1} \simeq \frac{1}{2\zeta} = \frac{\omega_0 m}{b}. \tag{6.37}$$

Finally, if the system is made to operate at the resonance frequency $\omega = \omega_0$, the amplitude $\hat{x}$ becomes

$$\hat{x} = \frac{F_0/m}{2\zeta\omega_0^2} = \frac{F_0}{2\zeta k} \simeq \frac{QF_0}{k}. \tag{6.38}$$

6.3.1   Example: Vibration Amplitude as a Function of the Damping Factor

Consider a forced vibration having a resonance frequency $\omega_0 = 2\pi \times 10^6$ rad/s. Plot the values of the amplitude $\hat{x}$, normalized to the low frequency value, as a function of the frequency for three values of the damping coefficient $\zeta = 1 \times 10^{-4}$, $\zeta = 2 \times 10^{-4}$ and $\zeta = 5 \times 10^{-4}$.

The equation is

$$X(s) = \frac{1}{s^2 + 2\zeta\omega_0 s + \omega_0^2}. \tag{6.39}$$

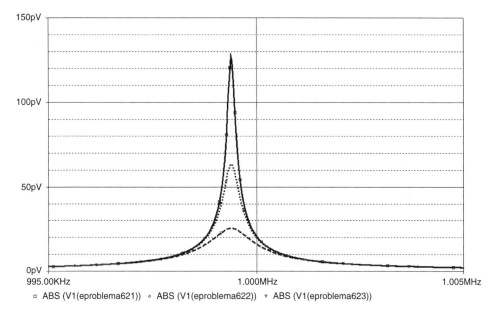

□ ABS (V1(eproblema621)) ◦ ABS (V1(eproblema622)) ▽ ABS (V1(eproblema623))

Figure 6.1 Plots of the amplitude normalized to the low frequency value for $\zeta = 1 \times 10^{-4}$ (solid curve), $\zeta = 2 \times 10^{-4}$ (dotted) and $\zeta = 5 \times 10^{-4}$ (dashed)

We write a PSpice file using a Laplace source:

```
vin 1 0 ac 1
eproblema621 11 0 laplace v(1)= 1/(s*(s+12.54e2+39.43e12/(s)))
eproblema622 12 0 laplace v(1)= 1/(s*(s+25.08e2+39.43e12/(s)))
eproblema623 13 0 laplace v(1)= 1/(s*(s+62.70 e2+39.43e12/(s)))
 .ac dec 100000 900e3 1100e3
 .probe
 .end
```

The results from running the simulation are shown in Figure 6.1. As can be seen, around the resonance frequency the values of the function are much larger than in rest of the spectrum.

6.4 Small Signal Equivalent Circuit of Resonators

Consider a lumped-parameter model of the mechanical system identical to the one described in Figure 5.9 when an electrostatic force F_E is applied:

$$m\frac{d^2x}{dt^2} + b\frac{dx}{dt} + kx = \frac{Q^2}{2\epsilon A} = F_E,$$ (6.40)

$$\frac{V_s - V}{R} = \frac{dQ}{dt},$$

where V_s is the source voltage and V is the voltage across the MEMS structure, R is the internal resistance of the source and x the displacement. Q is the charge stored at the capacitor. We suppose that the voltage has a bias DC component V_{SB} and a variable component or small signal v_S. The deflection will also have a steady-state component x_B, and a variable component x_S, hence $V_s = V_{SB} + \widetilde{V}_s$, $V = V_B + \widetilde{V}$ and $x = x_B + \widetilde{x}$. There are two variables, x and V. Linearizing F_E,

$$F_E(x, V) = F_E(x_B, V_B) + \left[\frac{\partial F_E}{\partial x}\right]_{x_B, V_B} (x - x_B) + \left[\frac{\partial F_E}{\partial V}\right]_{x_B, V_B} (V - V_B) \qquad (6.41)$$

and

$$F_E(x, V) = \frac{\epsilon_0 A V^2}{2(g_0 - x)^2} \approx \frac{\epsilon_0 A V_B^2}{2(g_0 - x_B)^2} + \frac{\epsilon_0 A V_B^2}{(g_0 - x_B)^3}\widetilde{x} + \frac{\epsilon_0 A V_B}{(g_0 - x_B)^2}\widetilde{V}. \qquad (6.42)$$

Substituting into equation (6.40),

$$m\frac{d^2\widetilde{x}}{dt^2} + b\frac{d\widetilde{x}}{dt} + \widetilde{x}\left[k - \frac{\epsilon_0 A V_B^2}{(g_0 - x_B)^3}\right] = \frac{\epsilon_0 A V_B^2}{2(g_0 - x_B)^2} - kx_B] + \frac{\epsilon_0 A V_B}{(g_0 - x_B)^2}\widetilde{V}. \qquad (6.43)$$

Taking into account that in steady state $\widetilde{x} = 0$ and $\widetilde{V} = 0$,

$$\frac{\epsilon_0 A V_B^2}{2(g_0 - x_B)^2} - kx_B = 0.$$

Then

$$m\frac{d^2\widetilde{x}}{dt^2} + b\frac{d\widetilde{x}}{dt} + \widetilde{x}\left[k - \frac{\epsilon_0 A V_B^2}{(g_0 - x_B)^3}\right] = \frac{\epsilon_0 A V_B}{(g_0 - x_B)^2}\widetilde{V}. \qquad (6.44)$$

Taking Laplace transforms,

$$ms^2\widetilde{X}(s) + bs\widetilde{X}(s) + \widetilde{X}(s)\left[k - \frac{\epsilon_0 A V_B^2}{(g_0 - x_B)^3}\right] = \frac{\epsilon_0 A V_B}{(g_0 - x_B)^2}\widetilde{V}(s). \qquad (6.45)$$

Linearizing the charge Q,

$$Q = \frac{\epsilon_0 A}{g_0 - x_B}(V_B + \widetilde{V})$$

$$= Q_B + \left[\frac{\partial Q}{\partial x}\right]_{x_B V_B} (x - x_B) + \left[\frac{\partial Q}{\partial V}\right]_{x_B V_B} (V - V_B)$$

$$= \frac{\epsilon_0 A}{g_0 - x_B}V_B + \frac{\epsilon_0 A}{(g_0 - x_B)^2}x_s + \frac{\epsilon_0 A}{g_0 - x_B}\widetilde{V}.$$

The circuit equation is given by

$$\frac{dQ}{dt} = \frac{\epsilon_0 A V_B}{(g_0 - x_B)^2}\frac{d\tilde{x}}{dt} + \frac{\epsilon_0 A V_B}{g_0 - x_B}\frac{dv_s}{dt} = \frac{V_{sB} + \tilde{V}_s - V}{R}. \tag{6.46}$$

Assuming that in steady state, $V_{SB} = V_B$, taking Laplace transforms,

$$\frac{\epsilon_0 A V_B}{(g_0 - x_B)^2}s\tilde{X}(s) + \frac{\epsilon_0 A}{g_0 - x_B}s\tilde{V}(s) = \frac{\tilde{V}_s(s) - \tilde{V}(s)}{R}. \tag{6.47}$$

From equations (6.45) and (6.47) it can be found that

$$\tilde{X}(s) = \tilde{V}(s)\frac{\dfrac{\epsilon A V_B}{(g_0 - x_B)^2}}{ms^2 + bs + k - \dfrac{\epsilon A V_B^2}{(g_0 - x_B)^3}} \tag{6.48}$$

and

$$\tilde{V}(s)\left(\frac{1}{R} + \frac{\epsilon A s}{g_0 - x_B} + \frac{s\dfrac{(\epsilon A V_B)^2}{(g_0 - x_B)^4}}{ms^2 + bs + k - \dfrac{\epsilon A V_B^2}{(g_0 - x_B)^3}}\right) = \frac{\tilde{V}_s(s)}{R}. \tag{6.49}$$

The Laplace transform of the current is

$$\tilde{I}(s) = \frac{\tilde{V}_s(s) - \tilde{V}(s)}{R}, \tag{6.50}$$

and the complex admittance $Y(s)$ of the resonator is

$$Y(s) = \frac{\tilde{I}(s)}{\tilde{V}(s)} = \frac{1}{R}\left(\frac{\tilde{V}_s(s)}{\tilde{V}(s)} - 1\right) \tag{6.51}$$

$$= s\frac{\epsilon_0 A}{g_0 - x_B} + \frac{1}{\dfrac{(g_0 - x_B)^4}{(\epsilon_0 A V_B)^2}\left(ms + b + \dfrac{1}{s}\left(k - \dfrac{\epsilon_0 A V_B^2}{(g_0 - x_B)^3}\right)\right)}. \tag{6.52}$$

From equation (6.52) the components of the admittance can be identified according to Figure 6.2. The inductance L is given by

$$L = m\frac{(g_0 - x_B)^4}{(\epsilon_0 A V_B)^2}, \tag{6.53}$$

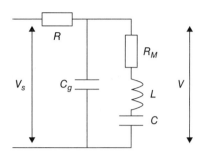

Figure 6.2 LCR equivalent circuit

the capacitance C by

$$C = \left(k \frac{(g_0 - x_B)^4}{(\epsilon_0 A V_B)^2} - \frac{g_0 - x_B}{\epsilon_0 A} \right)^{-1},$$ (6.54)

and the resistance R_M by

$$R_M = b \frac{(g_0 - x_B)^4}{(\epsilon_0 A V_B)^2}.$$ (6.55)

Of course there is an additional capacitive term in the admittance,

$$C_g = \frac{\epsilon_0 A}{g_0 - x_B},$$ (6.56)

and the internal source resistance R. The impedance of the equivalent circuit, neglecting the internal series resistance R, has two resonances, a series resonance of the LCR circuit, and the parallel resonance or anti-resonance. At the series resonance ω_s, the impedance of the series LCR arm is purely resistive and has the value R_M. Thus it is a low-impedance resonance. In contrast, the parallel resonance, ω_p, occurs when the circuit composed of the capacitor C_g and the two elements L and C reaches a maximum impedance value.

The series resonance frequency is given by

$$\omega_s = \frac{1}{\sqrt{LC}} = \sqrt{\frac{k}{m}} \sqrt{\frac{g_0 - 3x_B}{g_0 - x_B}}.$$ (6.57)

As can be seen, the series resonance frequency depends on the bias voltage as it depends on the bias displacement x_B, in such a way that when $x_B = 0$, $\omega_s = \sqrt{k/m}$. Of course when $x_B = g_0$ instability is reached and the result of equation (6.57) is infinity. The results in this section are only valid for small deflections around the equilibrium point where the small-signal approximation holds.

In contrast, the parallel resonance frequency occurs when the admittance of the circuit in Figure 6.2, neglecting R_M and R, is zero:

$$Y(j\omega_p) = j\omega C_g + \frac{1}{\frac{1}{j\omega C} + j\omega L} = 0,$$ (6.58)

and we have

$$\omega_p^2 = \frac{1}{LC_g}\left(1 + \frac{C_g}{C}\right).$$ (6.59)

Taking into account equation (6.57), the series and parallel resonances are related by

$$\omega_p^2 = \omega_s^2\left(1 + \frac{C}{C_g}\right).$$ (6.60)

6.4.1 Example: Series and Parallel Resonances

A resonator has the following equivalent circuit parameter values: $C_g = 1\,pF$, $C = 1\,fF$, $L = 1$ mH and $R_M = 10\,\Omega$. Carry out a PSpice simulation and plot the magnitude and phase of the impedance as a function of the frequency. Using the cursor utility, calculate the values of the series and parallel resonances.

A simple PSpice.cir file can be written for the impedance of the circuit with the given data values. The simulation frequency span has been adjusted to magnify the region of interest after a trial and error procedure.

```
vin 1 0 ac 1
eresonador 10 0 laplace v(1)= 1/(s*1e-12+ 1/(1e-3*s +1/(s*1e-15)+1e1))
.ac dec 100000 158.4e6 160e6
.probe
.end
```

Figure 6.3 shows the plot of the magnitude and phase. It can be seen that the modulus of the impedance has a minimum at a frequency of $f_s = \omega_s/2\pi = 159.155$ MHz (as found using the cursor utility in PSpice), which is the series resonance, and a maximum at a frequency of $f_p = \omega_s/2\pi = 159.235$ MHz which is the parallel resonance. The phase changes abruptly from $-\pi/2$ to $+\pi/2$ around the series resonance and from $+\pi/2$ to $-\pi/2$ around the parallel resonance. This corresponds to two resonances at the two points where the phase crosses 0. The parallel resonance is always greater than the series resonance, as shown in equation (6.60), and the phase is positive between the two resonances, indicating that the reactance is inductive between them.

6.4.2 Example: Spring Softening

Calculate the effective stiffness constant as a function of the bias deflection x_B.

The resonance frequency can be written as

$$\omega_s^2 = \frac{k_{eff}}{m} = \frac{k}{m}\frac{g_0 - 3x_B}{g_0 - x_B},$$ (6.61)

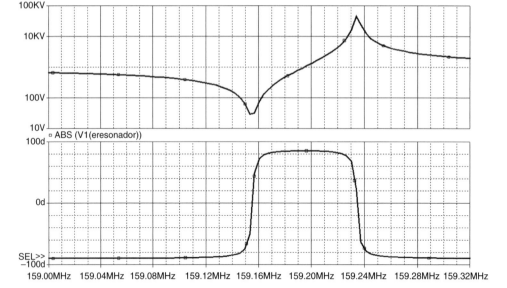

Figure 6.3 Series and parallel resonances: modulus of the impedance (top), and phase of the impedance (bottom)

and hence

$$k_{\text{eff}} = k \left(1 - \frac{2x_B}{g_0 - x_B} \right).$$ (6.62)

As can be seen, $k_{\text{eff}} < k$; the bigger the value of x_B, the smaller the value of k_{eff}.

6.5 Rayleigh–Ritz Method

In the section above we considered a point mass system, but in many cases in MEMS devices the mass is distributed and the lumped-element model fails. However, the energy method described above can be generalized [38]. In Section 3.10 we introduced Castigliano's method for finding the stiffness constant of flexures and defined the strain energy, that is, the system potential energy as the energy is elastically stored. Recall from equation (3.51) that

$$U_P = \int_V \frac{1}{2} \sigma_x \epsilon_x dx dy dz.$$ (6.63)

Taking into account that

$$\sigma_x = -\frac{Ey}{\rho_c} \quad \text{and} \quad \frac{1}{\rho_c} = \frac{d^2v}{dx^2},$$ (6.64)

where ρ_c is the radius of curvature and E is Young's modulus,

$$U_P = \int_V \frac{1}{2} E y^2 \frac{d^2 v}{dx^2} = \int \frac{1}{2} E \frac{d^2 v}{dx^2} dx \int y^2 dy dz = \int \frac{1}{2} EI \frac{d^2 v}{dx^2} dx. \qquad (6.65)$$

The kinetic energy is

$$U_K = \int_V \frac{1}{2} \rho_m \left(\frac{dv}{dt} \right)^2 dx dy dz, \qquad (6.66)$$

where ρ_m is the density. For a beam of rectangular section $dV = Whdx$, where W is the width and h the thickness,

$$U_K = \frac{1}{2} Wh \int \rho_m \left(\frac{dv}{dt} \right)^2 dx. \qquad (6.67)$$

In the vibration of a beam there are several vibration modes and the deflection can be modelled by a superposition of the several modes,

$$v(x, t) = \sum_n c_n w_n(x, t), \qquad (6.68)$$

where the $w_n(x, t)$ are known as shape functions and are defined as

$$w_n(x, t) = W_n(x) \sin(\omega_n t + \phi). \qquad (6.69)$$

The amplitude of a shape function is a function of the position only. The potential and kinetic energy for mode n are

$$U_{Pn\,max} = \frac{1}{2} EI c_n^2 \int \frac{d^2 W_n}{dx^2} dx \quad \text{and} \quad U_{Kn\,max} = \frac{1}{2} Whc_n^2 \omega_n^2 \int \rho_m W_n^2 dx. \qquad (6.70)$$

Applying the energy condition $U_{Pn\,max} = U_{Kn\,max}$,

$$\omega_n^2 = \frac{EI \int \left(\frac{d^2 W_n}{dx^2} \right)^2 dx}{Wh \int \rho_m W_n^2 dx}. \qquad (6.71)$$

If the shape functions are known, the vibration frequency of the different modes can be found as well. As a first approximation, the shape function can be assumed to be the steady-state solution of the deflection. This gives good results for the first harmonic, but not so good for the other harmonics.

6.5.1 Example: Vibration of a Cantilever

The cantilever shown in Figure 6.4 is subject to its own weight. Calculate the expression for the deflection as a function of the position x, and apply the Rayleigh–Ritz method to calculate the vibration frequency. Take $W = 5 \times 10^{-6}$ m, $h = 1 \times 10^{-6}$ m, $L = 100 \times 10^{-6}$ m, $g = 9.8$ m/s^2 and $\rho_m = 2329$ kg/m^3.

Assuming that the cantilever has a fixed end, the reaction at the support $R = WhL\rho_m g$ is equal to the total weight of the cantilever. Taking into account the free body diagram to the left of the position x and writing the force equilibrium, we obtain the force $V = R$. Taking moments around $x = 0$, and taking into account that the mass is distributed,

$$\int_0^x Wh\rho_m g\gamma d\gamma - M_b + M_R - Vx = 0, \tag{6.72}$$

where γ is an ancillary variable and

$$M_b = M_R + \rho_m g Wh\frac{x^2}{2} - Vx \tag{6.73}$$

is the bending moment. The beam equation is given by

$$EI\frac{d^2v}{dx^2} = M_R + \rho_m g Wh\frac{x^2}{2} - Vx \tag{6.74}$$

The boundary conditions are that $v(0) = 0$, $\frac{dv}{dx}|_{x=0} = 0$ and $\frac{d^2v}{dx^2}|_{x=L} = 0$. Using the last of these in equation (6.74), it follows that

$$M_R = \rho_m g Wh\frac{L^2}{2}. \tag{6.75}$$

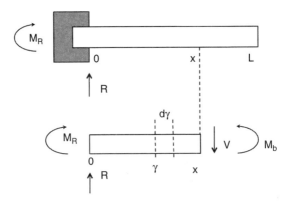

Figure 6.4 Example 6.5.1

The beam equation becomes

$$EI\frac{d^2v}{dx^2} = \frac{1}{2}\rho g WhL^2 + \rho_m g Wh\frac{x^2}{2} - \rho_m g WhLx. \tag{6.76}$$

Integrating twice and using the first two boundary conditions,

$$v = \frac{\rho_m g Wh}{24EI}x^2(x^2 + 6L^2 - 4Lx). \tag{6.77}$$

To make use of the Rayleigh–Ritz method to find the resonance frequency, we will assume that the first shape function is the result found in equation (6.77), $W_1 = v$, and hence

$$\frac{dW_1}{dx} = \frac{\rho_m g Wh}{24EI}(4x^3 + 12L^2x - 8Lx^2 - 4Lx^2) \tag{6.78}$$

and

$$\frac{d^2W_1}{dx^2} = \frac{\rho_m g Wh}{24EI}(12x^2 - 24Lx + 12L^2). \tag{6.79}$$

Using equation (6.79) in equation (6.71), it follows that the first mode is

$$\omega_1^2 = \frac{EI \int_0^L (12x^2 + 12L^2 - 24Lx)^2 dx}{\rho_m Wh \int_0^L x^4(x^2 + 6L^2 - 4Lx)^2 dx} = 1.03\frac{EI}{\rho_m WhL^4}. \tag{6.80}$$

6.5.2 Example: Gravimetric Chemical Sensor

We have a cantilever and we build a gravimetric sensor of volatile organic compounds based on the change in vibration frequency when the compounds are adsorbed and the cantilever mass changes. Calculate the expression relating the frequency change with the cantilever mass. Take $W = 5 \times 10^{-6}$ m, $h = 1 \times 10^{-6}$ m, $L = 100 \times 10^{-6}$ m, $g = 9.8$ m/s^2, $\rho = 2329$ kg/m^3.

For a cantilever with a rectangular cross-section, $I = Wh^3/12$. Taking into account that $m = \rho WhL$,

$$\omega_1^2 = 1.03\frac{EWh^3}{12mL^3}. \tag{6.81}$$

The sensitivity can be calculated from the derivative of the frequency with respect to the mass,

$$2\omega_1\frac{d\omega_1}{dm} = -1.03\frac{EWh^3}{12L^3}\frac{1}{m^2}, \tag{6.82}$$

and the sensitivity of the frequency to the mass is

$$\frac{d\omega_1}{dm} = -\frac{\omega_1}{2m}.$$ (6.83)

It can also be found that the relative change in frequency is related to the relative change in mass by

$$\frac{d\omega_1}{\omega_1} = -\frac{1}{2}\frac{dm}{m}.$$ (6.84)

For the data in this example the result will be a shift in the resonance frequency of 4.3% per picogram.

6.6   Resonant Gyroscope

Gyroscopes are devices which are able to measure angular velocity Ω. If we consider a point P in Figure 6.5(a) rotating at linear speed $\vec{v}$ at a position $\vec{r}$ away from the axis of rotation, the angular velocity is defined as a pseudo-vector $\overrightarrow{\Omega}$ such that $\vec{v} = \overrightarrow{\Omega} \times \vec{r}$, where $\times$ denotes the vector product.

In a gyroscope there are two coordinate axes: one is fixed and the other moves relative to the fixed one. The observation point also moves relative to the moving coordinate system.

The velocity of a given point relative to the fixed system of coordinates can be calculated as the superposition of the velocity of the same point relative to the movable coordinate system and of the linear velocity of the same point relative to the fixed coordinate system, when the movable coordinate system is at rest:

$$\left.\frac{dr}{dt}\right|_f = \left.\frac{dr}{dt}\right|_m + \Omega \times r.$$ (6.85)

If we now turn to the the geometry shown in Figure 6.5(b), the point can be identified by one position vector $\vec{r}$ relative to the movable coordinate system, and similarly by a second vector $\overrightarrow{r_0}$ relative to the fixed coordinate system. The two are related by

$$\overrightarrow{r_0} = \vec{R} + \vec{r}.$$ (6.86)

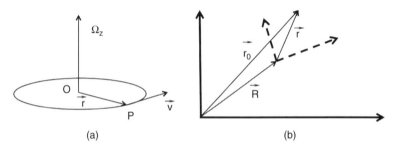

(a)   (b)

Figure 6.5   Definitions of angular movement: the fixed coordinate system is represented by the thick solid line and the moving system by the dashed line

The velocity of the point relative to the fixed system is then given by

$$\left.\frac{dr_0}{dt}\right|_f = \left.\frac{dR}{dt}\right|_f + \left.\frac{dr}{dt}\right|_f. \tag{6.87}$$

Taking into account equation (6.85),

$$\left.\frac{dr_0}{dt}\right|_f = \left.\frac{dR}{dt}\right|_f + \left.\frac{dr}{dt}\right|_m + \Omega \times r. \tag{6.88}$$

Taking the time derivative in equation (6.88),

$$\left.\frac{d^2r_0}{dt^2}\right|_f = \left.\frac{d^2R}{dt^2}\right|_f + \left.\frac{dv_m}{dt}\right|_f + \left.\frac{d\Omega}{dt}\right|_f \times r + \Omega \times \left.\frac{dr}{dt}\right|_f, \tag{6.89}$$

where

$$v_m = \left.\frac{dr}{dt}\right|_m. \tag{6.90}$$

Taking into account equation (6.85) and that a similar equation can be written for v_m,

$$\left.\frac{dv_m}{dt}\right|_f = \left.\frac{dv_m}{dt}\right|_m + \Omega \times v_m, \tag{6.91}$$

we have that

$$\left.\frac{d^2r_0}{dt^2}\right|_f = \left.\frac{d^2R}{dt^2}\right|_f + \left.\frac{dv_m}{dt}\right|_m + \Omega \times v_m + \left.\frac{d\Omega}{dt}\right|_f \times r + \Omega \times v_m + \Omega \times \left(\left.\frac{dr}{dt}\right|_m + \Omega \times r\right) \tag{6.92}$$

and

$$a_f = \left.\frac{d^2R}{dt^2}\right|_f + \left.\frac{dv_m}{dt}\right|_m + 2\Omega \times v_m + \left.\frac{d\Omega}{dt}\right|_f \times r + \Omega \times \Omega \times r. \tag{6.93}$$

In equation (6.93), the sum of the first two terms is the linear acceleration, the third term is the Coriolis acceleration, the fourth term is the local acceleration and the last term is the centripetal acceleration. If we simplify the problem into an angular velocity Ω having only a z-component Ω_z and neglecting the angular acceleration, $\Omega_y = \Omega_x = 0$ and $\left.\frac{d\Omega}{dt}\right|_f = 0$, the vector products can be written as

$$\Omega \times v_m = -\vec{i}\Omega_z v_{my} + \vec{j}\Omega_z v_{mx}, \tag{6.94}$$

$$\Omega \times r = -\vec{i}\Omega_z y + \vec{j}\Omega_z x, \tag{6.95}$$

$$\Omega \times \Omega \times r = -\vec{i}\Omega_z^2 x + \vec{j}\Omega_z^2 y, \tag{6.96}$$

where $\vec{i}$ and $\vec{j}$ are the unit vectors corresponding to the x- and y-axis, respectively. Neglecting second-order terms,

$$a_f = \frac{d^2 R}{dt^2}\bigg|_f + \frac{dv_m}{dt}\bigg|_m - \vec{i}2\Omega_z v_{my} + \vec{j}2\Omega_z v_{mx}. \tag{6.97}$$

Assuming that the linear acceleration of the movable coordinate system is zero, $\frac{d^2 R}{dt^2}\big|_f = 0$, and that the acceleration of the point relative to the movable coordinate system is restricted to two dimensions,

$$\frac{dv_m}{dt}\bigg|_m = \vec{i}\frac{d^2 x}{dt^2} + \vec{j}\frac{d^2 y}{dt^2}, \tag{6.98}$$

and as

$$v_{my} = \frac{dy}{dt} \quad \text{and} \quad v_{mx} = \frac{dx}{dt}, \tag{6.99}$$

we have

$$a_f = \vec{i}\left(\frac{d^2 x}{dt^2} - 2\Omega_z \frac{dy}{dt}\right) + \vec{j}\left(\frac{d^2 y}{dt^2} + 2\Omega_z \frac{dx}{dt}\right). \tag{6.100}$$

If we consider that we also may have external generic forces F_x and F_y along the x- and y-axes, and a moving mass m, the movement equations are

$$F_x - k_x x - b_x \frac{dx}{dt} = m_x \frac{d^2 x}{dt^2} - 2m_x \Omega_z \frac{dy}{dt}, \quad F_y - k_y y - b_y \frac{dy}{dt} = m_y \frac{d^2 y}{dt^2} + 2m_y \Omega_z \frac{dx}{dt}. \tag{6.101}$$

Usually one of the axes is used as driving electrode; for example, $F_x \neq 0$ and the y-axis is used as sensing electrode, hence $F_y = 0$. Equations (6.101) are a set of two coupled differential equations. Normally in MEMS applications, the movement of the drive element is much larger than the movement of the sense element, so $m_x(2\Omega \frac{dy}{dt})$ can be neglected. If the driving force is sinusoidal, $F_x = F_d \sin \omega t$, equations (6.101) simplify to

$$m_x \frac{d^2 x}{dt^2} + b_x \frac{dx}{dt} + k_x x = F_d \sin \omega t, \quad m_y \frac{d^2 y}{dt^2} + b_y \frac{dy}{dt} + k_y y = -2m_y \Omega_z \frac{dx}{dt}. \tag{6.102}$$

As can be seen, the first equation is independent of the variable y and can be solved separately, whereas the second equation depends on the variable x. Very importantly, the second term depends on the product of the angular velocity and the x-component of the linear velocity. So, in order to have sensitivity to the angular velocity it is indispensable to have the mass moving along the x-axis.

The resulting equation for the x-axis is the same as equation (6.15) in Section 6.3, where $\omega_x^2 = k_x/m_x$ Using feedback, this driving circuit is made to work at the resonant frequency $\omega = \omega_x$ and, according to (6.28),

$$\hat{x} = \frac{QF_d}{k_x}.$$ (6.103)

The Coriolis force is

$$-2m_y\Omega_z\frac{dx}{dt} = -2m_y\Omega_z\hat{x}\omega_x\cos(\omega_x t + \phi_x).$$ (6.104)

Then the y-axis differential equation is given by

$$m_y\frac{d^2y}{dt^2} + b_y\frac{dy}{dt} + k_y y = -2m\Omega_z\hat{x}\omega_x\cos(\omega_x t + \phi_x).$$ (6.105)

In resonance, if $\omega_x = \omega_y = \sqrt{k_y/m_y}$,

$$\hat{y} = \frac{2Q_y\hat{x}}{\omega_x}\Omega_z.$$ (6.106)

It can be concluded that the amplitude of the sensing displacement y is proportional to the displacement of the driving element and to the value of the angular velocity Ω_z.

6.7 Tuning Fork Gyroscope

Among the various ways to measure the angular velocity the one described in the section above based on the Coriolis acceleration can be implemented in MEMS technology using an architecture called the tuning fork [39]. This is shown in Figure 6.6.

The fixed parts involve a drive electrode (D) implemented by means of a comb drive that has half of its fingers attached to the fixed electrode D and half attached to the inertial mass

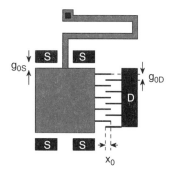

Figure 6.6 Tuning fork gyroscope

that is movable. There also are four sensing electrodes (S) that are fixed. The moving parts also involve a folded flexure in the upper part of the figure that is anchored in the little black square. The folded flexure usually is symmetric in the lower part of the figure but, it is omitted here for simplicity.

The comb structure drives the inertial mass into a movement along the x-axis, and an electronic control circuit ensures that it works at the resonant frequency of the x-axis. When there is an angular velocity Ω_z with a z-component out of the plane, the inertial mass undergoes a displacement along the y-axis, according to the equations developed in the previous section. This displacement is detected by the sense electrodes that initially, in the absence of angular velocity, are at a distance g_{0S} from the edge of the inertial mass. The gaps between the electrodes and the inertial mass change as the angular velocity changes, and the top gap and the bottom gap change in opposite directions: when the upper gap increases, the bottom gap decreases and vice versa. This is similar to the differential capacitor described in Section 5.13.

6.7.1   Example: Calculation of Sensitivity in a Tuning Fork Gyroscope

We have a gyroscope device such as the one depicted in Figure 6.6. The comb drive has 250 fingers. Calculate the expression for the electrostatic force assuming a DC voltage V_{CC} applied. If the drive voltage is changed to $V_{CC} + v_{AC}$ calculate the DC and AC force components at the drive frequency ω. Assuming resonant frequencies $\omega_x = \omega_y = \omega$, write the expression for the displacement on the x-axis and that on the y-axis as a function of the angular velocity Ω_z.

The capacitance of the comb drive is given by

$$C = 2N \frac{\epsilon_0 (x_0 + x) t_{comb}}{g_{0D}}, \tag{6.107}$$

where t_{comb} is the thickness of the comb drive fingers, g_{0D} is initial gap between the fingers in the comb drive and x_0 is the initial overlap of the comb drive fingers.

The force in the x-axis is

$$F_x = \frac{V_{CC}}{2} \frac{dC}{dx}\bigg|_V = \frac{N\epsilon_0 t_{comb}}{g_{0D}} V_{CC}^2. \tag{6.108}$$

If $V_{CC} + v_{AC}$ takes the place of V_{CC}, the force becomes

$$F_x = \frac{N\epsilon_0 t_{comb}}{g_{0D}} (V_{CC} + v_{AC})^2 = \frac{N\epsilon_0 t_{comb}}{g_{0D}} (V_{CC}^2 + v_{AC}^2 + 2V_{CC}v_{AC}). \tag{6.109}$$

The first term is a DC component of the force, the second is proportional to the signal v_{AC}^2 involving a $\sin^2(\omega t)$ which has a frequency twice ω and hence is far from the resonance frequency at which the system will operate. This term will be neglected, and the last term is the AC component of the force:

$$F_{xAC} = \frac{2N\epsilon_0 t_{comb}}{g_{0D}} V_{CC} v_d, \tag{6.110}$$

where v_d is the amplitude of the signal v_{AC}. As can be seen, the ac component of the force is proportional to the DC voltage applied.

The amplitude of the displacement along the x-axis is given by equation (6.103),

$$\hat{x} = \frac{Q_x F_{xAC}}{k_x} = \frac{2N\epsilon_0 t_{comb} Q_x}{g_{0D} k_x} V_{CC} v_d, \tag{6.111}$$

and we have

$$\hat{y} = \frac{2Q_y \hat{x}}{\omega_x} \Omega = \frac{4N\epsilon_0 t_{comb} Q_x Q_y}{g_{0D} k_x \omega_x} V_{CC} v_d \Omega. \tag{6.112}$$

The sensitivity is

$$S = \frac{\hat{y}}{\Omega} = \frac{4N\epsilon_0 t_{comb} Q_x Q_y}{g_{0D} k_x \omega_x} V_{CC} v_d. \tag{6.113}$$

Problems

6.1 Show that the factor Q is equal to the resonant frequency divided by the bandpass.

6.2 With the same data values as in Example 6.3.2, calculate the value of factor Q from the results of the PSpice simulation.

6.3 We have a vibrating mass of a plate $500 \times 500\ \mu m^2$ made of silicon with a density of $2339\ kg/m^3$, and the plate thickness is $h = 0.2\ \mu m$. The initial gap is $g_0 = 1\ \mu m$. We assume that the applied bias voltage makes the plate deflect by $x_B = \frac{1}{4} g_0$, corresponding to a bias voltage of 10 V. We assume a damping coefficient of $b = 3 \times 10^{-3}$. Calculate the values of the components of the equivalent LCR circuit.

6.4 A laterally resonating gravimetric sensor is described in [40]. The proof mass is made out of the patterning of the 5 μm device layer of a silicon on insulator wafer. The other two dimensions of the mass are 160 μm and 30 μm. The density of silicon is $2339\ kg/m^3$. From the reference, the resonant frequency can be identified as $\omega_0 = 245.5\ kHz$. (1) Calculate the value of the effective stiffness constant of the flexures supporting the proof mass. (2) Calculate the value of frequency shift that can be expected if a polystyrene bead of 550 pg is deposited on top of the proof mass.

6.5 Calculate the value of the free vibration frequency, the value of the damped vibration frequency and the frequency at which the maximum amplitude of deflection occurs for a cantilever with the following data: $W = 10 \times 10^{-6}$ m, $h = 1.5 \times 10^{-6}$ m, $L = 200 \times 10^{-6}$ m, $g = 9.8$ m/s^2, $\rho = 2329$ kg/m^3, $E = 130$ GPa, $\zeta = 0.1$.

6.6 In a tuning fork gyroscope similar to the one described in Section 6.7, the comb drive is composed of 250 fingers and the main parameter values are: $t_{comb} = 2$ μm, $g_{0D} = 1$ μm, $x_0 = 100$ μm, $\omega_x = 2\pi \times 15 \times 10^3$ rad/s, $V_{CC} = 30$ V, $Q_x = Q_y = 10\,000$. Calculate the sensitivity.

6.7 Calculate the effective stiffness of a resonant cantilever due to the steady-state electrostatic attraction as a function of the bias position of the cantilever x_B. The dimensions of the cantilever are $500 \times 10 \times 2$ μm, the initial gap is $g_0 = 5$ μm and the elastic stiffness $k = 1$ N/m.

6.8 A double-clamped bridge has a first resonance given by [19, p. 218]

$$\omega(0) = 22.27 \sqrt{\frac{EI}{\rho_m W t L^4}}. \tag{6.114}$$

When the bridge is subject to an axial force, the resonance frequency shifts due to the axial force F, as follows [41]:

$$\omega(F) = \omega(0) \sqrt{1 + \gamma \frac{FL^2}{12EI}}, \tag{6.115}$$

where $\gamma = 0.2949$. If the beam is subject to a pressure $P = 10$ kPa, calculate the shift in resonance frequency. The bridge has a length of $L = 200$ μm, a width $W = 20$ μm and thickness $t = 2$ μm. ρ_m is the density of the bridge material, assumed to be silicon: $\rho_m = 2340$ kg/m^3. Young's modulus is $E = 169$ GPa.

6.9 A resonant accelerometer is based on a proof mass attached to two double-ended tuning fork resonators [42]. Find the expected frequency change for an acceleration of $10g$. The dimensions of the beam are $L = 450$ μm, $W = 6$ μm and $t = 10$ μm. Young's modulus is $E = 169$ GPa. The resonant frequency in the absence of acceleration is $f(0) = 81.978$ kHz.

7

Microfluidics and Electrokinetics

There are many phenomena in microscopic dimensions related to fluids where the viscous forces dominate, making the hydrodynamic pressure inefficient for moving fluids. There are a number of electrokinetic forces:

- the electrothermal force acting on the volume of the fluid due to temperature gradients;
- the electro-osmotic force acting near the boundary surfaces;
- the electrowetting force acting on the triple line (fluid–solid–gas interface);
- the electrophoretic and dielectrophoretic forces acting on embedded particles in a fluid.

In this chapter we describe the basics of the fluid transport by pressure difference and by the electrokinetic force that can be applied due to the electrical double layer for ionic liquids. We concentrate also on the electrowetting principle that is gaining momentum in applications such as displays and lab-on-chip devices. We conclude with a description of the dielectrophoretic force that is used to drive and sort nanoparticles.

7.1 Viscous Flow

Let us consider a differential volume in a pipe of rectangular cross-section as shown in Figure 7.1. The change with respect to time of the mass inside the volume can be written as [43]

$$\frac{\partial}{\partial t}(\rho_m \, h \, dxdy). \tag{7.1}$$

If we suppose that the fluid has a velocity u in the direction of the x-axis, the volume entering at x per unit time is

$$\rho_m \, dy \, u \, dz, \tag{7.2}$$

Understanding MEMS: Principles and Applications, First Edition. Luis Castañer.
© 2016 John Wiley & Sons, Ltd. Published 2016 by John Wiley & Sons, Ltd.
Companion Website: www.wiley.com/go/castaner/understandingmems

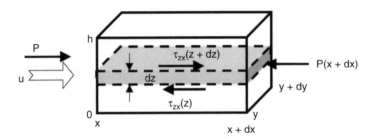

Figure 7.1 Differential of volume in a pipe

where ρ_m is the fluid mass density.[1] Extending to the z-axis,

$$\rho_m \, dy \int u \, dz, \tag{7.3}$$

we now introduce the flow in the x-direction by unit transversal length,

$$q(x) = \int u \, dz. \tag{7.4}$$

The rate of change of mass inside the volume is equal to the difference between the volume entering at x and the volume leaving at $x + dx$,

$$\frac{\partial}{\partial t}(\rho_m \, h \, dx \, dy) = \rho_m \, dy \, q(x) - \rho_m \, dy \, q(x + dx). \tag{7.5}$$

Extending to two dimensions,

$$\frac{\partial}{\partial t}(\rho_m \, h) = \rho \, dy \, q(x) - \rho_m \, dy \, q(x + dx) + \rho_m \, dx \, q(y) - \rho_m \, dx \, q(y + dy). \tag{7.6}$$

leading to

$$\frac{\partial \rho_m \, h}{\partial t} + \frac{\partial}{\partial x}(\rho_m \, q(x)) + \frac{\partial}{\partial y}(\rho_m \, q(y)) = 0. \tag{7.7}$$

We set the equilibrium conditions in the volume by considering the pressure, $P(x)$ and $P(x + dx)$, at the two horizontal ends and the shear forces at the bottom and top surfaces, $\tau_{zx}(z)$ and $\tau_{zx}(z + dz)$,

$$P(x)dydz + \tau_{zx}(z + dz)dxdy = P(x + dx)dydz + \tau_{zx}(z)dxdy. \tag{7.8}$$

---

[1] Note that in this chapter ρ_m is the density of fluid, not to be confused with the radius of curvature ρ_c in previous chapters.

We can write

$$\frac{\partial \tau_{zx}}{\partial x} = \frac{\partial P}{\partial x}. \tag{7.9}$$

The shear force τ_{zx} can be substituted by its definition as a function of the viscosity, μ, and the fluid velocity in the slice u_z,

$$\tau_{zx} = \mu \frac{\partial u_z}{\partial z}. \tag{7.10}$$

Viscosity can be interpreted in terms of the shear forces between different layers of a fluid moving at different speeds. τ_{zx} is acting along the x-axis on a surface whose normal is along the z-axis. Equation (7.10) indicates that the shear forces are proportional to the gradient of the fluid velocity in the transversal direction. Substituting equation (7.10) into equation (7.8),

$$\frac{\partial P}{\partial x} = \frac{\partial}{\partial z} \left(\mu \frac{\partial u}{\partial z} \right) \tag{7.11}$$

One particular case is when the pressure gradient is not a function of z, and this is called Poiseuille flow. Integrating equation (7.11) twice and applying boundary conditions of zero velocity at the pipe walls, $u(0) = 0$ and $u(h) = 0$, the fluid velocity is found to be

$$u(z) = \frac{1}{2\mu} \frac{\partial P}{\partial x} z(z - h). \tag{7.12}$$

This result indicates that the velocity profile in the the z-direction is quadratic in z and the maximum velocity occurs when

$$\frac{\partial u_x(z)}{\partial z} = 0, \quad \frac{1}{2\mu} \frac{\partial P}{\partial x} ((z - h) + z) = 0. \tag{7.13}$$

The solution of equation (7.13) shows that the maximum of the velocity occurs at the centre of the pipe at $z = h/2$, and the value of the maximum velocity is

$$u_{x\,max} = -\frac{h^2}{8\mu} \frac{\partial P}{\partial x}. \tag{7.14}$$

In this Poiseuille flow case, if we suppose that the pipe has width W in the transversal direction to the flow and that the velocity distribution does not depend on y, the value of the flow, F, is given by

$$F = \int_0^W \int_0^h u_x(z) dy dz = \frac{W}{2\mu} \frac{\partial P}{\partial x} \int_0^h (z^2 - zh) dz = -\frac{W h^3}{12\mu} \frac{\partial P}{\partial x} \tag{7.15}$$

where the minus sign indicates that the flow is in the direction of decreasing pressure.

More generally, if the flow is two-dimensional we can calculate the flows per unit transversal length. In the x-direction,

$$q(x) = \int_0^h u(z)dz = -\frac{h^3}{12\mu}\frac{\partial P}{\partial x}, \tag{7.16}$$

and in the y-direction,

$$q(y) = \int_0^h u(z)dz = -\frac{h^3}{12\mu}\frac{\partial P}{\partial y}. \tag{7.17}$$

Substituting equation (7.16) and equation (7.17) into equation (7.6),

$$\frac{\partial}{\partial t}(\rho_m h) = \frac{\partial}{\partial x}\left(\frac{h^3}{12\mu}\frac{\partial P}{\partial x}\right) + \frac{\partial}{\partial y}\left(\frac{h^3}{12\mu}\frac{\partial P}{\partial y}\right). \tag{7.18}$$

In an isothermal fluid the pressure and density are proportional, $\rho_m/\rho_{m0} = P/P_0$, and taking into account that

$$P\frac{\partial P}{\partial x} = \frac{1}{2}\frac{\partial P^2}{\partial x}, \quad P\frac{\partial P}{\partial y} = \frac{1}{2}\frac{\partial P^2}{\partial y}, \tag{7.19}$$

it follows that

$$\nabla^2 P^2 = \frac{24\mu}{h^3}\frac{\partial}{\partial t}(Ph), \tag{7.20}$$

which is the Reynolds equation.

7.2 Flow in a Cylindrical Pipe

The flow in a cylindrical pipe can be calculated by transforming equation (7.11) to cylindrical coordinates,

$$\frac{\partial P}{\partial x} = \mu\frac{1}{r}\frac{\partial}{\partial r}\left(r\frac{\partial u}{\partial r}\right). \tag{7.21}$$

Integrating equation (7.21) twice and applying the boundary condition that the velocity is zero at the wall ($r = a$) and that there is symmetry at the centre ($r = 0$),

$$u(r = a) = 0, \quad \frac{du}{dr}\Big)_{r=0} = 0, \tag{7.22}$$

it follows that

$$r\frac{du}{dr} = \frac{1}{\mu}\int\frac{\partial P}{\partial x}rdr = \frac{1}{\mu}\left(\int\frac{\partial P}{\partial x}\frac{r^2}{2} + C_1\right). \tag{7.23}$$

Using the symmetry boundary condition, it follows that $C_1 = 0$. Integrating a second time,

$$u = \frac{r^2}{4\mu}\frac{\partial P}{\partial x} + C_2,$$ (7.24)

and using the wall boundary condition,

$$C_2 = -\frac{a^2}{4\mu}\frac{\partial P}{\partial x}$$

and, finally,

$$u = \frac{a^2}{4\mu}\frac{\partial P}{\partial x}(r^2 - a^2).$$ (7.25)

Again the flow has direction opposite to the pressure gradient. As the differential cross-section is $2\pi\, r\, dr$, the flow across a section of the pipe is given by

$$F = \int_0^a u(r)2\pi rdr = \int_0^a 2\pi rdr\frac{a^2}{4\mu}\frac{\partial P}{\partial x}(r^2 - a^2)dr = -\frac{\pi}{8\mu}a^4\frac{\partial P}{\partial x}.$$ (7.26)

7.2.1   Example: Pressure Gradient Required to Sustain a Flow

We want to deliver a 10 μl droplet in 0.1 seconds through a cylindric channel of length $L = 1\,mm$ and radius $a = 0.1\,mm$ (see Figure 7.2). The liquid has a viscosity of $\mu = 8.9 \times 10^{-4}$ Pa s. Calculate the pressure difference that must be applied to the channel.

The flow F is

$$F = 10 \times 10^{-6} \times \frac{1\,\text{m}^3}{10^3}\frac{1}{0.1\,\text{s}} = 10^{-7}\,\text{m}^3/\text{s}.$$ (7.27)

Using equation (7.26),

$$F = -\frac{\pi a^4}{8\mu}\frac{\partial P}{\partial x} = -\frac{\pi}{8 \times 8.9 \times 10^{-4}}(0.1 \times 10^{-3})^4\frac{\Delta P}{L}$$ (7.28)

and

$$\Delta P = 2.26\,\text{kPa}.$$ (7.29)

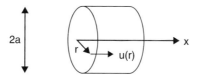

Figure 7.2   Geometry for Example 6.2.1

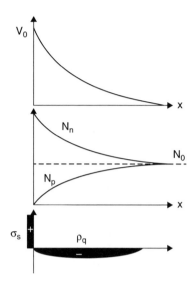

Figure 7.3   Potential, ion concentrations and charge distribution from the electrode-solution interface

7.3   Electrical Double Layer

Consider a plane surface carrying a uniform positive charge density in contact with a solution having both positive and negative ions. There is a distribution of the concentration of ions from the interface to deep inside the solution that corresponds to the electrical potential distribution from a value at the surface to a reference voltage far away from the interface. The simplest model is the Guoy–Chapman model based on the probability of finding an ion of valence z at a distance from the interface where the potential is V [44]; this probability follows a Boltzmann distribution. Figure 7.3 shows the concentrations of positive, N_p, and negative, N_n, ions per unit volume,

$$N_p = N_0 e^{-qzV/kT}, \quad N_n = N_0 e^{qzV/kT}, \tag{7.30}$$

such that as the potential tends to zero, far from the surface, both concentrations tend to the same value N_0. q is the electron charge. The net charge density per unit volume, ρ_q (measured in C/m^3),[2] can be written as

$$\rho_q = zq(N_p - N_n) = zqN_0(e^{-qzV/kT} - e^{qzV/kT}) = -2zqN_0 \sinh \frac{qzV}{kT}. \tag{7.31}$$

The region close to the interface with the electrode where we have this net charge distribution is called the diffuse layer. The charge neutrality establishes that the total excess charge inside the solution per unit transversal area must be equal to the surface charge density at the electrode,

$$\sigma_s = -\int \rho_q dx. \tag{7.32}$$

---

[2] Not to be confused with the mass density ρ_m.

The Poisson equation can be written

$$\nabla^2 V = -\frac{\rho_q}{\epsilon} \tag{7.33}$$

$$\simeq -\frac{2zq}{\epsilon} \sinh \frac{qzV}{kT} N_0. \tag{7.34}$$

An analytical solution of equation (7.34) was found by Guoy and Chapman which, for values of the voltage small compared to kT/q, can be simplified to

$$\nabla^2 V = -\frac{2z^2 q^2 N_0}{\epsilon kT} V; \tag{7.35}$$

this is known as the Debye–Hückel equation.

The Debye length, λ_D, is defined as

$$\lambda_D^2 = \frac{\epsilon kT}{2z^2 q^2 N_0}; \tag{7.36}$$

at 25°C, and for an ion solution at concentration C (mol/m^{-3}), taking into account that N_0 (m^{-3}) $= CN_A$, where N_A is the Avogadro number, can be written as

$$\lambda_D = \frac{9.61 \times 10^{-9}}{z C^{1/2}}. \tag{7.37}$$

The Debye–Hückel equation can be written as

$$\nabla^2 V = -\frac{V}{\lambda_D^2} \tag{7.38}$$

The solution, subject to the boundary conditions that we have a potential at the surface $V = V_0$ at $x = 0$ and that the potential deep into the solution and far from the surface is $V = 0$ at $x \longrightarrow \infty$, is

$$V = V_0 e^{-x/\lambda_D}. \tag{7.39}$$

A relationship between the surface charge σ_s and the potential is easily found from the charge neutrality condition in equation (7.32)

$$\sigma_s = -\int_0^\infty \rho_q dx = \epsilon \int_0^\infty \nabla^2 V dx = \epsilon \left[\left(\frac{dV}{dx} \right)_\infty - \left(\frac{dV}{dx} \right)_0 \right] = -\epsilon \left(\frac{dV}{dx} \right)_0, \tag{7.40}$$

the final equality following from

$$\left(\frac{dV}{dx} \right)_\infty = 0. \tag{7.41}$$

Equation (7.40) yields

$$\sigma_s = \frac{\epsilon V_0}{\lambda_D} \tag{7.42}$$

This can be interpreted as the relationship between the surface charge of a unit-area parallel-plate capacitor where the plates are separated by λ_D with a dielectric in between with permittivity ϵ. As can be seen, if the temperature is low, the solution concentration is high, the charge number carried by the ions is high and the dielectric constant is also high, then the diffuse layer can be very thin, creating a compact layer of ions just separated from the electrode by a distance the size of the solvent molecules. This is called the Helmholtz layer. More detailed models such as the Stern–Grahame model considers that there are two Helmholtz planes, an inner one and an outer one, and that the potential drop splits between the two. An electrical double layer is the basis of supercapacitors.

7.3.1 Example: Debye Length and Surface Charge

For a monovalent 0.01 M solution of a 1:1 electrolyte in water, we know that the surface potential is 25 mV. Calculate the Debye length and the surface charge.

Using equation (7.37), and assuming for the solution a relative permittivity value of $\epsilon_r = 80$ F/m, and that $z = 1$, it follows that

$$\lambda_D = \frac{9.61 \times 10^9}{zC^{1/2}} = \frac{9.61 \times 10^9}{0.01^{1/2}} = 96.1 \text{ nm}. \tag{7.43}$$

The surface charge is

$$\sigma_s = \frac{\epsilon V_0}{\lambda_D} = \frac{80 \times 8.82 \times 10^{-12} \times 25 \times 10^{-3}}{96.1 \times 10^{-9}} = 1.84 \times 10^{-4} \text{ C/m}^2. \tag{7.44}$$

7.4 Electro-osmotic Flow

In ionic liquids the double layer that forms at the walls may help the flow of the liquid from the input inlet to the output. This can be achieved by applying an electric field between inlet and outlet by means of suitable electrodes. The equation governing such flow, called electro-osmotic flow, is the simplified Navier–Stokes equation,

$$\mu \frac{d^2u}{dx^2} = -\rho_q \mathbf{E}, \tag{7.45}$$

where ρ_q is the charge density per unit volume and $\mathbf{E}$ is the applied electric field between two electrodes located along the Y-axis. Equation (7.45) relates the second derivative of the flow velocity in the X-direction to the applied electric field in the Y-direction. Taking into account equations (7.34) and (7.38),

$$\rho_q = \frac{\epsilon V}{\lambda_D^2}. \tag{7.46}$$

Using equation (7.39),

$$\rho_q = \frac{\epsilon V_0 e^{-x/\lambda_D}}{\lambda_D^2}, \tag{7.47}$$

and taking into account equation (7.42),

$$\rho_q = \frac{\sigma_s e^{-x/\lambda_D}}{\lambda_D^2}. \tag{7.48}$$

Equation (7.45) can now be written

$$\frac{d^2 u}{dx^2} = \frac{\sigma_s \mathbf{E}}{\mu \lambda_D} e^{-x/\lambda_D}. \tag{7.49}$$

Equation (7.49) can be integrated with two boundary conditions:

$$u(x = 0) = 0 \quad \text{and} \quad \left. \frac{du}{dx} \right|_{x \to \infty} = 0. \tag{7.50}$$

The first boundary condition simply sets the flow velocity to zero at the wall and the second sets a zero gradient of velocity far from the wall. The first integration of (7.49) gives

$$\frac{du}{dx} = -\frac{\sigma_s \mathbf{E}}{\mu} e^{-x/\lambda_D} + C_1 \tag{7.51}$$

Applying the boundary condition in the first derivative, it follows that $C_1 = 0$. Then the second integration of (7.49) gives

$$u = \frac{\sigma_s \mathbf{E} \lambda_D}{\mu} e^{-x/\lambda_D} + C_2. \tag{7.52}$$

Applying the boundary condition on the velocity at $x = 0$, $C_2 = -\sigma_s \mathbf{E} \lambda_D / \mu$, and the velocity is

$$u = -\frac{\sigma_s \mathbf{E} \lambda_D}{\mu} (1 - e^{-x/\lambda_D}). \tag{7.53}$$

As can be seen, the maximum of the velocity when $x \gg \lambda_D$, is

$$u_{\max} = -\frac{\sigma_s \mathbf{E} \lambda_D}{\mu},$$ (7.54)

and with equation (7.42),

$$u_{\max} = \frac{\epsilon V_0 \mathbf{E} \lambda_D}{\mu}.$$ (7.55)

As the Debye length is generally very small compared to the other dimensions, the maximum of the velocity is reached very close to the wall and hence, for channels where the width is much greater than the value of the Debye length, the velocity can be assumed constant throughout the cross-section and equal to the maximum given by equation (7.55). The electro-osmotic flow in the case of a rectangular pipe of width W and height h, is then given by

$$F = \int u_{\max} dA = u_{\max} Wh = \frac{\epsilon V_0 \mathbf{E} \lambda_D}{\mu} Wh.$$ (7.56)

7.5 Electrowetting

Electrowetting is the name given to the phenomenon observed by Gabriel Lippmann on the behaviour of a drop of conductive liquid on top of a substrate when an electrical potential is applied between the drop and the substrate [45]. The shape of the drop depends on the surface tension; for a hydrophobic surface the contact angle is greater than 90°. When a voltage is applied, the contact angle is reduced and the liquid wets the substrate more. This phenomenon has not found practical applications until recently when a thin dielectric was deposited covering the substrate making the effect more important and reliable [46]. This is called electrowetting on dielectric.

We can use the definition of surface tension to analyse the air–liquid–substrate system depicted in Figure 7.4. A battery V_s is connected by means of a needle to a drop of liquid that

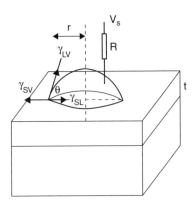

Figure 7.4 Electrowetting definitions

we assume to be electrically conductive. The substrate is also electrically conductive. Between liquid and substrate there is thin layer of dielectric material of thickness t. Let us consider two situations: before and after the voltage is applied. There are several components of the energy change in this battery–droplet system: the energy of the solid–air interface, ΔE_{SV}; the energy of the solid–liquid interface, ΔE_{SL}; the energy of the liquid–air interface, ΔE_{LV}; the energy of the battery, ΔE_B; and the energy of the capacitor, ΔE_C, where ΔE_C comes from the fact that at the bottom interface of the drop of liquid we have a capacitor composed of two electrodes (the liquid itself and the substrate) and the dielectric in between. From conservation of energy we have

$$\Delta E_{SV} + \Delta E_{SL} + \Delta E_{LV} + \Delta E_B + \Delta E_C = 0. \tag{7.57}$$

From Figure 7.4 an increase dA in the liquid–solid interface area produces a decrease $-dA$ in the solid–air surface area and an increase of $dA \cos \theta$ in the liquid–vapour surface area:

$$\Delta E_{SV} = -\gamma_{SV} dA, \quad \Delta E_{SL} = \gamma_{SL} dA, \quad \Delta E_{LV} = \gamma_{LV} \cos \theta dA. \tag{7.58}$$

The change in the battery energy is given by

$$\Delta E_B = \frac{dE_B}{dA} dA = \frac{V_s dQ_B}{dA} dA = V_s \frac{dQ_B}{dA} dA, \tag{7.59}$$

where the voltage drop in the resistor R (assumed to be the internal resistance of voltage source) has been neglected and Q_B is the charge stored in the battery. From the circuit, it is evident that the charge that goes out of the battery dQ_B has to go into the capacitor dQ_C, hence $dQ_B = -dQ_C$. As $Q_C = CV$,

$$\Delta E_B = -\frac{V dQ_C}{dA} dA = -V \frac{d(CV)}{dA} dA = -V^2 \frac{dC}{dA} dA = -V^2 \frac{\epsilon}{t} dA \tag{7.60}$$

as the voltage is kept constant and the differential of capacitance is $dC = \frac{\epsilon}{t} dA$.

The energy in a capacitor is given by $\frac{1}{2} CV^2$ and hence the capacitor energy change is given by

$$\Delta E_C = \frac{dE_C}{dA} dA = \frac{V^2}{2} \frac{dC}{dA} dA = \frac{V^2}{2} \frac{\epsilon}{t} dA. \tag{7.61}$$

Returning to equation (7.57), it follows that

$$-\gamma_{SV} dA + \gamma_{SL} dA + \gamma_{LV} \cos \theta dA + \frac{V^2 \epsilon}{2t} dA - \frac{V^2 \epsilon}{t} dA = 0. \tag{7.62}$$

Hence,

$$-\gamma_{SV} + \gamma_{SL} + \gamma_{LV} \cos \theta - \frac{V^2 \epsilon}{2t} = 0. \tag{7.63}$$

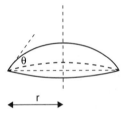

Figure 7.5   Spherical cap geometry of a droplet

C_A is defined as the ratio ϵ/t which is the capacitance per unit area. Equation (7.63) can be written as

$$- \gamma_{SV} + \gamma_{SL} + \gamma_{LV} \cos\theta - \frac{V^2 C_A}{2} = 0. \tag{7.64}$$

This is the Young–Dupré equation relating the surface and interface tensions with the contact angle and with the term $V^2 C_A/2$ which is the energy stored in the capacitance per unit area.

Equation (7.64) can also be written in terms of the contact angle θ_0 between the liquid and solid before the voltage is applied, so for $V = 0$, equation (7.64) leads to

$$- \gamma_{SV} + \gamma_{SL} + \gamma_{LV} \cos\theta_0 = 0, \quad \text{and hence} \quad \cos\theta_0 = \frac{\gamma_{SV} - \gamma_{SL}}{\gamma_{LV}}. \tag{7.65}$$

Finally, substituting equation (7.65) into equation (7.64), it follows that

$$\cos\theta = \cos\theta_0 + \frac{C_A V^2}{2\gamma_{LV}} = 0, \tag{7.66}$$

thus providing the relationship between the contact angle values before and after applying the voltage.

Analytical models frequently use a simplification of the shape of a drop on top of a surface by considering a spherical cap shape. This is shown in Figure 7.5. If the volume of the drop (*vol*) is known, the radius of the base of the drop for a contact angle θ is

$$r = \left(\frac{3vol}{\pi}\right)^{1/3} \frac{\sin\theta}{(2 - 3\cos\theta + \cos^3\theta)^{1/3}}. \tag{7.67}$$

7.5.1   Example: Droplet Change by Electrowetting

We know that a 20 μl drop of water on top of a surface covered by a t = 80 nm thick dielectric layer has a contact angle of θ = 110°. The solid–air surface tension is $\gamma_{SV} = 19.1 \times 10^{-3}$ N/m and the water–air surface tension is $\gamma_{LV} = 72.8 \times 10^{-3}$ N/m. Calculate the solid–water surface tension γ_{SL}. What is the value of the radius of the base of the drop, assuming a spherical cap geometry? What is the voltage required to reduce the contact angle to θ = 75°, assuming that *the relative permittivity of the dielectric is $\epsilon_r = 7.5$? What is the value of the radius of the base of the drop after applying such voltage?*

Using equation (7.66),

$$\cos\theta_0 = \frac{\gamma_{SV} - \gamma_{SL}}{\gamma_{LV}} = \frac{19.1 \times 10^{-3} - \gamma_{SL}}{72.8 \times 10^{-3}} = \cos 110 = -0.342. \tag{7.68}$$

The solid–liquid surface tension is then $\gamma_{SL} = 43.9 \times 10^{-3}$ N/m.
Using equation (7.67), the radius is

$$r = \left(\frac{3 \times 10^{-9}}{3.14}\right)^{1/3} \frac{\sin 110}{(2 - 3\cos 110 + \cos^3 110)^{1/3}} = 1.73 \times 10^{-3} \text{ m}. \tag{7.69}$$

Using equation (7.66) and taking into account that $C_A = \epsilon_0 \epsilon_r / t$,

$$\cos\theta = \cos\theta_0 + \frac{C_A V^2}{2\gamma_{LV}}. \tag{7.70}$$

Substitution yields

$$\cos 75 = \cos 110 + \frac{7.5 \times 8.85 \times 10^{-12} V^2}{2 \times 72.8 \times 10^{-3} \times 80 \times 10^{-9}}, \tag{7.71}$$

whence $V = 10.6$ V. Using equation (7.67) again for $\theta = 75°$, it follows that $r = 1.82 \times 10^{-3}$ m.

7.5.2   Example: Full Substrate Contacts

Electrowetting devices can be driven using the two electrodes in the bottom substrate to avoid the need for upper electrodes, whether needles or electrode covered plates. Figure 7.6 shows the cross-section of an appropriate device. One of the electrodes is grounded while the other is the driving electrode. The two electrodes are underneath the thin dielectric layer as shown [47, 48]. (a) Draw an electric equivalent circuit of the coupling of the electrodes. (b) Calculate the voltage effectively applied between the liquid and the driving electrode. (c) Calculate the change in contact angle when a voltage $V = 10$ V is applied between the driving electrode and ground. (d) If a contact angle change of $10°$ is required, what should the driving electrode voltage be? Use the following data. The radius of the grounded electrode is $R_G = 0.2$ mm. The radius of the outer circle of the driving electrode is $R_D = 0.5$mm. The gap between the grounded electrode and the inner circle of the driving electrode is $g = 0.05$ mm. The thickness of the dielectric is $t = 200$ nm. Finally, $\gamma_{LV} = 72.8 \times 10^{-3}$ N/m and the initial contact angle $\theta_0 = 110°$.

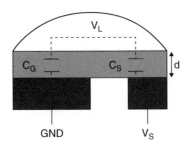

Figure 7.6 Example 7.5.2

(a) As shown in Figure 7.6, there are two capacitances: one between the liquid and the driving electrode, C_D; and the other between the grounded electrode and the liquid, C_G. As the driving voltage is applied between the driving electrode and ground, there is a capacitive voltage divider where the middle point is the liquid itself.

(b) As the charge is the same in both series capacitors,

$$(V_D - V_L)C_D = V_L C_G \quad \text{and} \quad V_L = V_D \frac{C_D}{C_D + C_G}. \tag{7.72}$$

As

$$C_D = \frac{\epsilon_0 \epsilon_r A_D}{d} \quad \text{and} \quad C_G = \frac{\epsilon_0 \epsilon_r A_G}{d}, \tag{7.73}$$

we have

$$V_L = V_D \frac{A_D}{A_D + A_G} \quad \text{and} \quad V_D - V_L = V_D \frac{A_G}{A_D + A_G} = 1.76 \text{ V}. \tag{7.74}$$

(c) The change in contact angle is found by applying equation (7.66),

$$\cos \theta = \cos \theta_0 + \frac{\epsilon_0 \epsilon_r (V_D - V_L)^2}{2 \gamma_{LV}} = -0.3395, \tag{7.75}$$

so the contact angle is $\theta = 109.85°$. From this result it can be learned that the capacitive coupling significantly reduces the voltage effectively applied between the driving electrode and the liquid, thereby reducing the contact angle change.

(d) For 10° change, $\theta = 100°$,

$$V_D - V_L = \sqrt{\frac{2t\gamma_{LV}(\cos \theta - \cos \theta_0)}{\epsilon_0 \epsilon_r}} = 14.9 \text{ V}, \tag{7.76}$$

and $V_D = 84.9$ V.

7.6 Electrowetting Dynamics

A simplified model to describe the dynamics of the triple line has been derived assuming the the drop has a spherical cap geometry [49]. We consider the circuit shown in Figure 7.4. The starting point is the Young–Dupré equation (7.64) in equilibrium and in the absence of any applied voltage,

$$-\gamma_{SV} + \gamma_{SL} + \gamma_{LV} \cos\theta = 0. \tag{7.77}$$

Since we want a dynamic model, we must include in the equilibrium equation a friction term γ_F per unit length which we will assume to be proportional to the triple-line velocity v,

$$-\gamma_{SV} + \gamma_{SL} + \gamma_{LV} \cos\theta + \gamma_F = 0. \tag{7.78}$$

We will further suppose that out of equilibrium, a change in the stored electrical energy U_A in the solid–liquid capacitance C_A per unit area leads to a differential change in the solid–liquid surface energy per unit area (i.e. surface tension),

$$d\gamma_{SL} = -dU_A. \tag{7.79}$$

Taking the time derivative of equation (7.77),

$$\frac{d\gamma_{SV}}{dt} = -\gamma_{LV}\sin\theta\frac{d\theta}{dt} + \frac{d\gamma_{SL}}{dt} + \frac{d\gamma_F}{dt} = 0, \tag{7.80}$$

since we assume that both γ_{LV} and γ_{SV} do not depend on the voltage. From equation (7.79), we can write

$$dU_A = C_A V dV = q_A dV, \tag{7.81}$$

where V is the voltage drop across the liquid and the substrate, assumed to be grounded, and q_A is the stored charge per unit area. As

$$d\gamma_{SL} = \frac{d\gamma_{SL}}{dt}dt = -q_A\frac{dV}{dt}dt, \tag{7.82}$$

and considering that $\gamma_F = \xi v$,

$$-\gamma_{LV}\sin\theta\frac{d\theta}{dt} + \xi\frac{dv}{dt} - q_A\frac{dV}{dt} = 0. \tag{7.83}$$

As the triple-line velocity $v = dr/dt$ and the radius of the bottom circle of the drop is given by equation (7.67),

$$v = \frac{dr}{dt} = -\alpha\Omega(\theta)\frac{d\theta}{dt}, \tag{7.84}$$

where

$$\Omega(\theta) = \frac{(1 - \cos\theta)^2}{(2 - 3\cos\theta + \cos^3\theta)^{3/4}} \quad \text{and} \quad \alpha = \left(\frac{3vol}{\pi}\right)^{1/3}, \qquad (7.85)$$

we find the differential equation

$$-\xi\alpha\Omega(\theta)\frac{d^2\theta}{dt^2} - \xi\alpha\left(\frac{d\Omega(\theta)}{dt}\right)^2 - \gamma_{LV}\sin\theta\frac{d\theta}{dt} - \frac{q_A}{C_A}\frac{dq_A}{dt} = 0. \qquad (7.86)$$

Furthermore, we can write a circuit equation from the circuit shown in Figure 7.4:

$$V_s = R\frac{dq}{dt} + \frac{q}{C}. \qquad (7.87)$$

As the area of the base circle of the drop is given by

$$A = \frac{\pi\alpha^2\sin^2\theta}{(2 - 3\cos\theta + \cos^3\theta)^{2/3}}, \qquad (7.88)$$

equation (7.87) becomes

$$\frac{dq_A}{dt} = \frac{V_s}{AR} - \frac{q_A}{A}\left(\frac{dA}{d\theta}\frac{d\theta}{dt} + \frac{1}{C_AR}\right). \qquad (7.89)$$

Moreover,

$$\frac{dA}{d\theta} = \frac{dA}{dr}\frac{dr}{d\theta} = 2\pi r\frac{dr}{d\theta}, \qquad (7.90)$$

and for the spherical cap geometry,

$$\frac{dr}{d\theta} = -\alpha\Omega, \qquad (7.91)$$

hence

$$\frac{dA}{d\theta} = -2\pi r\alpha\Omega, \qquad (7.92)$$

together with equation (7.89), leads to

$$\frac{dq_A}{dt} = \frac{V_s}{AR} - q_A\left(-\frac{2\alpha\Omega}{r}\frac{d\theta}{dt} + \frac{1}{AC_AR}\right). \qquad (7.93)$$

Equations (7.86) and (7.93) are a system of of differential equations that can be solved for the values of the contact angle θ as a function of time.

7.6.1 Example: Contact-angle Dynamics

Write Matlab code to solve equations (7.86) and (7.93) for the following data values, which are representative of a water droplet on top of a substrate covered by a thin layer of Teflon. Find and plot the values of the contact angle and of the triple-line velocity as a function of time, when the initial value of the contact angle is $\theta_0 = 100°$, the droplet volume is 5×10^{-9} l, $V_s = 70$ V, $\xi = 20$, $\gamma_{LV} = 52.65 \times 10^{-3}$, $\gamma_{SV} = 1.4 \times 10^{-3}$, $d = 0.9$ μm, $\epsilon_0 = 8.85 \times 10^{-12}$, $\epsilon_d = 2.1$ and $R = 1$ Ω.

Matlab code to solve these equations is given in the solution to Problem 7.8. The results are shown in Figure 7.7. As can be seen, the change in contact angle occurs roughly 1 second after the voltage is applied. The contact angle change is in the range of 1 radian.

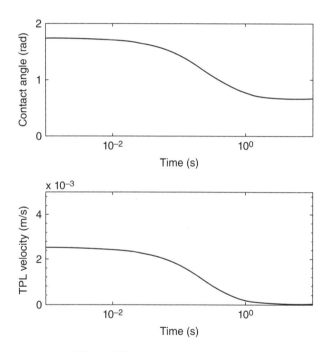

Figure 7.7   Results for Example 7.6.1

7.7 Dielectrophoresis

In many applications, nanoparticles have to be moved, separated, focused etc. Many such particles are electrically neutral and, when placed in an electric field, do not respond to the Coulomb force but become polarized and behave as electrical dipoles.

7.7.1 Electric Potential Created by a Constant Electric Field

Assume that there are no charges, that the charges creating the constant electric field $\mathbf{E}_0$ are far away and that the field has the direction of the z-axis.

The electric potential can be calculated as

$$V = V_1 + V_2 - \int_O^M \overrightarrow{\mathbf{E}dl}, \tag{7.94}$$

where we consider the point O as the origin of coordinates and $\overrightarrow{dl}$ is the differential of length along the integration path. The integral in equation (7.94) is independent of the path of integration because the potential satisfies the Laplace equation as the electric field is divergence-free and curl-free. If we choose the path of integration as ON and NM (Figure 7.8), it follows that

$$V = - \int_O^N \overrightarrow{\mathbf{E}dl} - \int_N^M \overrightarrow{\mathbf{E}dl}. \tag{7.95}$$

As the electric field is directed along the z-axis, the first integral in equation (7.95) is zero, and in the second integral

$$\mathbf{E} = \vec{k}E_0 \quad \text{and} \quad \overrightarrow{dl} = \vec{k}(M - N) = \vec{k}z. \tag{7.96}$$

Hence

$$\overrightarrow{\mathbf{E}dl} = E_0 z = E_0 r \cos\theta \tag{7.97}$$

and

$$V = -E_0 r \cos\theta. \tag{7.98}$$

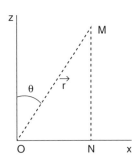

Figure 7.8 Geometry for calculating the potential created by a constant electric field directed along the z-axis

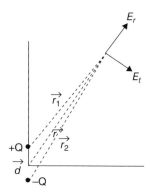

Figure 7.9 Definitions of radial and tangential components of electric field

7.7.2 Potential Created by an Electrical Dipole

Suppose that the centre of the dipole is located at the centre of coordinates and that the dipole has a charge $+Q$ separated by a distance d from a $-Q$ charge (Figure 7.9).

The potential is

$$V = \frac{Q}{4\pi\epsilon r_1} - \frac{Q}{4\pi\epsilon r_2} = \frac{Q}{4\pi\epsilon r_1}\frac{r_2 - r_1}{r_1 r_2}. \tag{7.99}$$

This can be approximated when the distance $r \gg d$ by

$$r_1 = r - \frac{d}{2}\cos\theta \quad \text{and} \quad r_2 = r + \frac{d}{2}\cos\theta \tag{7.100}$$

and

$$V = \frac{Q}{4\pi\epsilon}\frac{d\cos\theta}{r^2}. \tag{7.101}$$

From this equation the radial and tangential components of the electric field created by the dipole at the point M can be calculated:

$$\mathbf{E}_r = -\frac{\partial V}{\partial r} = \frac{Q}{2\pi\epsilon}\frac{d\cos\theta}{r^3} \tag{7.102}$$

and

$$\mathbf{E}_t = \frac{\partial V}{\partial r} = \frac{Q}{4\pi\epsilon}\frac{d\sin\theta}{r^3}. \tag{7.103}$$

The quantity $Qd = p_0$ is defined as the dipole moment.

7.7.3 Superposition

Consider the case where there is a constant electric field directed along the z-axis simultaneously with a dielectric sphere of radius R having its centre at the same origin of coordinates. The potential at a given point M has two components, one originated by the electric field $\mathbf{E}_0$ and the other originated by the electric dipole due to the polarization of the spherical dielectric particle. We will suppose that the total potential has the form of the superposition of the two components described above [6, p. 10]:

$$V_{\text{out}} = Ar\cos\theta + B\frac{\cos\theta}{r^2} \quad \text{and} \quad V_{\text{in}} = Cr\cos\theta + D\frac{\cos\theta}{r^2}, \tag{7.104}$$

where A, B, C and D are constants calculated from boundary conditions. It is required that the potential at $r \to \infty$ be the potential created by the constant electric field E_0 as the other terms vanish. This means that $A = -E_0$. Furthermore, the potential when $r \to 0$ should not tend to infinity, hence $D = 0$, and we have

$$V_{\text{out}} = -\mathbf{E}_0 r\cos\theta + B\frac{\cos\theta}{r^2} \quad \text{and} \quad V_{\text{in}} = Cr\cos\theta. \tag{7.105}$$

The two additional boundary conditions required to calculate the constants B and C are, first, that the potential inside and outside is the same at the sphere surface as there is no charge, and second, the displacement flux vector $\overrightarrow{D}$ is continuous at the surface, so at $r = R$,

$$V_{\text{out}} = V_{\text{in}} \quad \text{and} \quad \epsilon_1 \frac{\partial V_{\text{out}}}{\partial r} = \epsilon_2 \frac{\partial V_{\text{in}}}{\partial r}, \tag{7.106}$$

where ϵ_1 is the permittivity of the medium and ϵ_2 is the permittivity of the particle. The constants and B and C are calculated as

$$B = \frac{\epsilon_2 - \epsilon_1}{\epsilon_2 + 2\epsilon_1} R^3 E_0, \quad C = \frac{3\epsilon_1 - \epsilon_1}{\epsilon_2 + 2\epsilon_1} E_0. \tag{7.107}$$

Comparing the result in equation (7.98) with the dipole-induced voltage outside the sphere given in the second term of V_{out} in equation (7.105), we obtain

$$V = \frac{p_{\text{eff}}}{4\pi\epsilon} \frac{\cos\theta}{r^2}, \tag{7.108}$$

where

$$p_{\text{eff}} = 4\pi\epsilon_1 K R^3 E_0, \quad K = \frac{\epsilon_2 - \epsilon_1}{\epsilon_2 + 2\epsilon_1}. \tag{7.109}$$

These equations describe the potential created by a sphere of diameter R polarized by an electric field constant in the direction of the z-axis. K is the Clausius–Mossotti factor. When K is positive the effective moment $\overrightarrow{p_{\text{eff}}}$ is parallel to the electric field $\overrightarrow{E_0}$ and if K is negative

It is antiparallel. With this expression for p_{eff} the dielectrophoretic force is calculated using equation (1.48),

$$\vec{F} = 2\pi\epsilon_1 R^3 K \nabla |\mathbf{E}|^2,$$

which is equation (1.50). The force has the same direction as the gradient of the square of the modulus of the electric field.

Problems

7.1 A drop of water has a contact angle on top of a surface of $\theta_0 = 110°$. We know that $\gamma_{SV} = 10.1 \times 10^{-3}$ N/m and $\gamma_{LV} = 72.8 \times 10^{-3}$ N/m. The bottom conductive substrate is covered by a 80 nm thick layer of Si_3N_4 having a relative permittivity of $\epsilon_r = 7.5$. Calculate the value of γ_{SL}. Calculate the radius of the base circle of a droplet of 20 μl, assuming that the shape can be approximated by a spherical cap. Calculate the voltage that must be applied for the contact angle to have a value of 75°. Calculate the new value of the radius of the base circle.

7.2 A drop of water on top of a substrate forms a contact angle of 110° and when a 10 V voltage is applied, the contact angle reduces to 90°. Calculate the surface tension γ_{LV} for a capacitance per unit area of $C_A = 1 \times 10^{-4}$ F/m^2.

7.3 One way to avoid direct contact on the liquid using a needle or an upper electrode is to make use of a capacitive coupling between electrodes in the bottom substrate as shown in Figure 7.6. The dielectric has a thickness $d = 200$ nm, $\gamma_{LV} = 72.8 \times 10^{-3}$, the initial contact angle is $\theta_0 = 110°$, and $\epsilon_r = 2.1$. Write the relationship between the voltages V_S and V_L. Calculate the value of the contact angle after a voltage $V_S = 10$ V is applied.

7.4 With the same data as in Problem 7.3, we would like to change the design of the device in order to make the change in contact angle larger. For this reason we reduce the thickness of the dielectric to $d = 50$ nm. Calculate the ratio of the two areas A_S/A_G to achieve a change of 10° in the contact angle after $V_S = 10$ V is applied.

7.5 We would like to compare the electro-osmotic flow with pressure-driven flow. For the same flow in both cases, calculate the correspondence between the electric field value and the pressure gradient value for the following data: $V_0 = 100$ mV, $W = 100$ μm, $h = 20$ μm, $\epsilon_0 = 8.85 \times 10^{-12}$ F/m and $\epsilon_r = 80$.

7.6 For the same data as in Problem 7.5, calculate the pressure gradient required to sustain a flow of 200 μl/min in a PDMS channel 1000 μm long. The liquid viscosity is 0.000749 Pa s.

7.7 We have a 100×100 μm cross-section channel of length $L = 1$ mm. We want to deliver a flow of 100 μl/hour. Calculate the pressure difference that must be applied. The liquid viscosity is $\mu = 8.9 \times 10^{-4}$ Pa s.

7.8 Write Matlab code to solve the electrowetting equations (7.86) and (7.93).

7.9 Calculate the contact angle from maximum to minimum after applying a 70 V voltage using the same data as in Example 7.6.1, but for several values of the friction coefficient $\zeta = 1, 20, 50$. Calculate approximately from the plots the time required for the change.

7.10 Calculate the change in contact angle for the same data as in Example 7.7.1, but for several values of the applied voltage, $V = 20, 40, 60, 75$ V.

7.11 Dielectrophoresis is used to levitate, classify and sort nanoparticles. In [50] a series of interdigitated electrodes are used to make latex particles levitate. Calculate the required value of the Clausius–Mossotti factor for a particle of 3 μm radius suspended in water to counterbalance the gravitational force, if the estimated value of the electric field gradient squared is 10^{12} V^2/m^3.

8

Thermal Devices

A significant number of physical properties of matter depend on temperature, and many MEMS applications are based on thermal phenomena. In this chapter the basics of heat equations are covered, concentrating on the heat conduction and convection transfer processes. Applications to thermal flow sensors, thermocouples and thermal bimorph actuators are described.

8.1 Steady-state Heat Equation

Heat transfer can be produced (a) by thermal conduction between two parts of a body at different temperatures, (b) by convection between a body and a fluid in contact with it, and (c) by exchange of energy by electromagnetic radiation between a body and its environment. Let us start with heat conduction [51]. Fourier's law establishes that the heat flow inside a thermal conducting body is proportional to the temperature gradient and is directed from higher temperature to lower temperature. Taking into account the definitions in Figure 8.1, the heat flux, q_{cond},[1] which has units of W/m^2, at a given plane inside a body having a distribution of temperature $T(x)$, is given by

$$q_{cond} = -k\frac{dT}{dx},\tag{8.1}$$

where k^2 is the thermal conductivity with units of W/m $\cdot$ K, and T is the temperature. The minus sign in equation (8.1) tells us that the heat flux goes in the opposite direction to the temperature gradient.

The convection of heat is the transfer from the surface of a body at a given temperature T_S to a fluid that has a temperature far from the body surface T_∞. The convection heat flux, q_{conv}, is given by

$$q_{conv} = h(T_S - T_\infty)\tag{8.2}$$

where h is the heat convection transfer coefficient and has units of W/m$^2 \cdot$ K.

---

[1] The symbol q is used in this chapter for heat flux, and is not to be confused with liquid flow in Chapter 7 and with the electron charge.

[2] k in this chapter is thermal conductivity not to be confused with the stiffness constant used in previous chapters.

Understanding MEMS: Principles and Applications, First Edition. Luis Castañer.
© 2016 John Wiley & Sons, Ltd. Published 2016 by John Wiley & Sons, Ltd.
Companion Website: www.wiley.com/go/castaner/understandingmems

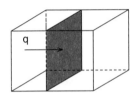

Figure 8.1   Heat conduction

Radiation is the exchange of energy between a heated surface and its environment in the form of electromagnetic radiation. There is no need for a medium between the emitting and absorbing bodies, as was necessary in conduction and convection, as the transfer is produced by photons. A common situation is when a small surface at a given temperature T_1 is surrounded by another surface at a different temperature T_2. The heat radiation from the surface is given

$$q_{\text{rad}} = \epsilon_m \sigma_{\text{SB}}(T_1^4 - T_2^4), \tag{8.3}$$

where ϵ_m is the body emissivity and σ_{SB} is the Stefan–Boltzmann constant.

If, through the differential volume shown in Figure 8.2 and neglecting radiation, there is heat flow from the input flux q_{in} to the output flux q_{out}, eventually there will be heat generated in the volume, q_{gen}, and heat transfer due to convection, q_{conv}. At equilibrium,

$$\dot{Q}_{\text{in}} + \dot{Q}_{\text{gen}} = \dot{Q}_{\text{out}} + \dot{Q}_{\text{conv}}, \tag{8.4}$$

where $\dot{Q}$ is the heat rate measured in watts and is given by the integral of the heat flux q at a surface A,

$$\dot{Q} = \int_A q\, dA \tag{8.5}$$

Taking into account Fourier's law of heat conduction,

$$q_{\text{in}} = -k \left.\frac{dT}{dx}\right|_x \quad \text{and} \quad q_{\text{out}} = -k \left.\frac{dT}{dx}\right|_{x+\Delta x}, \tag{8.6}$$

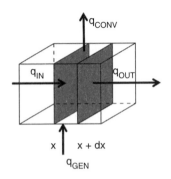

Figure 8.2   Heat equation components

and that the convection term is given by

$$q_{conv} = h\Delta T, \tag{8.7}$$

where h is the convection coefficient, the equilibrium equation can be written as

$$-k\left.\frac{dT}{dx}\right|_x A_{cond} + \dot{Q}_{gen} = -k\left.\frac{dT}{dx}\right|_{x+\Delta x} A_{cond} + hA_{conv}\Delta T, \tag{8.8}$$

where A_{cond} is the cross-section area through which the heat conduction occurs and A_{conv} is the area of the body in contact with the fluid. We can also write

$$\left.\frac{dT}{dx}\right|_{x+\Delta x} = \left.\frac{dT}{dx}\right|_x + \left.\frac{d^2T}{dx^2}\right|_x \Delta x. \tag{8.9}$$

Substituting in equation (8.8),

$$k\left.\frac{d^2T}{dx^2}\right|_x A_{cond}\Delta x = -\dot{Q}_{gen} + hA_{conv}\Delta T. \tag{8.10}$$

The differential equation (8.10) allows us to find the temperature distribution $T(x)$ in this one-dimensional geometry subject to boundary conditions.

8.2 Thermal Resistance

The heat equation can be simplified to generate a lumped-parameter equivalent model. Let us consider the case where we only have heat conduction. The heat equation (8.8) establishes that the heat rate entering the volume is equal to the heat rate exiting the volume; thus, using equation (8.5),

$$\dot{Q} = -k\int_A \frac{dT}{dx} dA \simeq -kA\frac{\Delta T}{\Delta x}. \tag{8.11}$$

If the cross-section A is constant, we can write

$$\dot{Q} = -\frac{\Delta T}{\Theta}, \tag{8.12}$$

where

$$\Theta = \frac{\Delta x}{kA} \tag{8.13}$$

is called the thermal resistance between the two points separated by a distance Δx and having a cross-section A in between. If heat is conducted by a body of low thermal conductance, the thermal resistance will be large and, for a given value of the heat rate, the temperature difference between the to ends of the body will be large as well. Similarly, for a given thermal

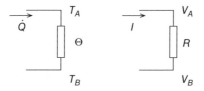

Figure 8.3   Thermal equivalent circuit compared with an electrical circuit

conductivity value, a body of smaller cross-section leads to a large temperature difference. High thermal resistance means high thermal isolation. If we have two points A and B in the body, the heat rate flowing from A to B, following equation (8.12), is written

$$\dot{Q} = -\frac{T_B - T_A}{\Theta} = \frac{T_A - T_B}{\Theta}, \tag{8.14}$$

where $T_A > T_B$.

Equation (8.14) is often interpreted in terms of an electrical equivalent circuit, where the heat rate has the meaning of an electrical current, the drop in temperature has the meaning of a voltage drop and the thermal resistance has the meaning of an electrical resistance, as shown in Figure 8.3.

The thermal resistance concept can also be extended to convection, as a convection thermal resistance Θ_{conv} can also be defined,

$$\dot{Q}_{\mathrm{conv}} = h\Delta T A_{\mathrm{conv}}, \tag{8.15}$$

and then

$$\Theta_{\mathrm{conv}} = \frac{1}{hA_{\mathrm{conv}}}. \tag{8.16}$$

8.2.1   Example: Temperature Profile in a Heated Wire

If the geometry is that of a wire and if the wire is made of a resistive material that is heated by an electrical current I, calculate expressions for A_{cond}, A_{conv} and q_{gen}. Calculate an expression for dR assuming that the material of the wire has a resistivity ρ. For parameter values thermal conductivity $k = 401$ W/m $\cdot$ K, $d = 10^{-6}$ m, $I = 100$ mA, $T_{\mathrm{air}} = 283$ K, $T_0 = 283$ K, $L = 10^{-3}$ m, $\rho = 1.68 \times 10^{-8}\ \Omega$ m and and $T_{\mathrm{ref}} = 283$, calculate and plot the temperature profile along the wire, assuming that the temperature at the two ends of the wire is T_0. Solve it for three values of the convection coefficient $h = 5, 25, 100$ W/m^3 $\cdot$ K.

From Figure 8.4, the conduction and convection areas are

$$A_{\mathrm{cond}} = \pi\frac{d^2}{4}, \quad A_{\mathrm{conv}} = \pi d\Delta x, \tag{8.17}$$

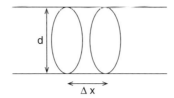

Figure 8.4 Geometry for Example 8.3.1

where d is the wire diameter and Δx is the segment of the wire considered. The current I through the wire dissipates a power in the differential resistor dR. This is the heat flux generated within the volume of the segment of the wire,

$$\dot{Q}_{gen} = I^2 dR. \qquad (8.18)$$

By the definition of resistance (see equation (4.101)), and if the cylindrical body we are considering is made of a resistive material having an electrical resistivity ρ,

$$dR = \frac{\rho \Delta x}{\pi d^2/4}. \qquad (8.19)$$

Equation (8.10) becomes

$$\left. \frac{d^2 T}{dx^2} \right|_x = \frac{4h\Delta T}{kd} - \frac{16 I^2 \rho}{k\pi^2 d^4}. \qquad (8.20)$$

To solve this equation in Matlab, the vector of the function and its first derivative is

$$\begin{pmatrix} y(1) \\ y(2) \end{pmatrix} = \begin{pmatrix} T \\ \frac{dT}{dx} \end{pmatrix}, \qquad (8.21)$$

and the vector of the derivatives of the vector y, called $dydx$, is given by

$$dydx = \begin{pmatrix} dydx(1) \\ dydx(2) \end{pmatrix} = \begin{pmatrix} \frac{dT}{dx} \\ \frac{d^2 T}{dx^2} \end{pmatrix} = \begin{pmatrix} y(2) \\ \frac{4h\Delta T}{kd} - \frac{16 I^2 \rho}{k\pi^2 d^4} \end{pmatrix}. \qquad (8.22)$$

The solution of equation (8.20) gives the temperature profile in the wire along the x-axis provided the boundary conditions at the two ends of the wire are known (in this case we have T_0). The core of the Matlab code is as follows:

```
solinit= bvpinit(linspace(0,L,5),[To 0]);
sol=bvp4c(@hotwirefunctionbook,@hotwirebcbook,solinit);
x=linspace(0,L);
y=deval(sol,x);
```

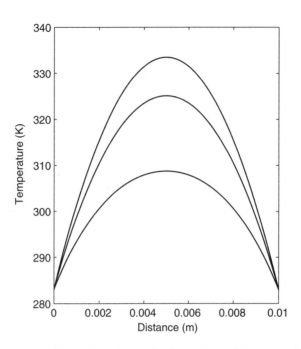

Figure 8.5 Temperature profile in a heated wire for three values of the convection coefficient $h = 5$ (top), 25 (middle) and 100 (bottom) W/m$^2 \cdot$ K

Here we have used a boundary value problem solver (bvp4c) with an initial solution $T = T_0$ along the wire, equation (8.20) is written in the 'hotwirefunctionbook.m' file and the boundary conditions in the 'hotwirebcbook.m' file:

```
function dydx=hotwirefunctionbook(x,y)
dydx=[y(2)
4*h*(y(1)-Tair)/(k*d)-(I^2*resistivity)/(k*pi^2*d^4)];
```

The boundary conditions are written as:

```
function res= hotwirebcbook(ya,yb)
res=[ya(1)-To
yb(1)-To];
```

The solution of the code for the parameter values above is shown in Figure 8.5. As can be seen, at the two ends the equilibrium temperature holds, and there is a temperature distribution along the wire that has a maximum at the mid-point due to symmetry. When the convection coefficient h is low the heat lost to convection is small and the maximum overheat at the mid-point reaches $\Delta T \simeq 50$ K. However, when the heat convection coefficient becomes larger the overheat takes smaller values.

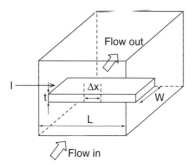

Figure 8.6 Heated resistor in a channel

8.2.2   Example: Resistor Suspended in a Bridge

The use of resistors suspended in a channel allows us to manufacture thermal flow sensors. A fluid is made to flow through the channel and, somewhere in the middle, there is a micro-machined bridge which can be heated electrically by means of an electric current as shown in Figure 8.6.

The flow refrigerates the suspended device and the average temperature changes accordingly. This method can be used to measure flow as described in Section 8.5 below. We suppose that the bridge is suspended at the two ends in the direction orthogonal to the flow and an electrical current is also fed by those two points. The heat generation is created by Joule dissipation in the resistor and there are heat conduction losses through the supports. Moreover, the fluid flow refrigerates the volume. The heat conduction cross-section is Wt, while the convection area is composed of the two horizontal surfaces and two vertical surfaces opposing the fluid flow; therefore by the definitions in Figure 8.6, The main parameters are:

$$A_{\text{cond}} = Wt, \quad A_{\text{conv}} = 2\Delta x(W + t) \tag{8.23}$$

and

$$dR = \frac{\rho \Delta x}{Wt}. \tag{8.24}$$

Equation (8.8) becomes

$$\left.\frac{d^2 T}{dx^2}\right|_x = \frac{2h(W + t)\Delta T}{kWt} - \frac{I^2 \rho}{k(Wt)^2}. \tag{8.25}$$

8.3 Platinum Resistors

Platinum resistors are very well known in the electronics industry because the temperature coefficient of the resistivity of platinum is positive and the linearity extends over a wide range of temperatures. Platinum resistors are usually modelled by the relationship [52]

$$R(T) = R_0(1 + AT + BT^2), \tag{8.26}$$

where R_0 is the value of the resistor at $T = 0°C$ and, in the range $0 < T < 850°C$, $A = 3.908 \times 10^{-3}°C^{-1}$ and $B = -5.775 \times 10^{-7}°C^{-2}$. The relative temperature coefficient α of commercial platinum resistors is defined as

$$\alpha = \frac{R_{100} - R_0}{100R_0} = A + 100B \approx A, \tag{8.27}$$

where R_{100} is the value of the resistor at $T = 100°C$. In most applications the value of α is taken as $\alpha \simeq A$, neglecting the quadratic term. Platinum can be deposited as a thin film on top of a substrate. If the current I in Figure 8.6 is constant, the voltage generated across the resistor is

$$V = IR \simeq IR_0(1 + \alpha T). \tag{8.28}$$

Platinum resistors can also be used to dissipate power like any other conventional resistor, but then the power dissipated will be a function of the temperature,

$$P = I^2 R \simeq I^2 R_0(1 + \alpha T). \tag{8.29}$$

These resistors are used as temperature sensors and also as heaters, as shown below.

8.4 Flow Measurement Based on Thermal Sensors

The principle of thermal anemometry is based on the change in the convection transfer of heat to the fluid, which is a function of the flow velocity. The various methods used involve the calorimetric approach and involve placing three resistors along the flow, heating the middle one and sensing the temperature difference between the resistors upstream and downstream, as schematically shown in Figure 8.7. Another approach is the anemometric method which involves placing a heater and a temperature sensor in a channel either in one of the walls or suspended. The heater receives electric power and an equilibrium temperature is established.

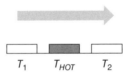

Figure 8.7 Calorimetric measurement of flow

The fluid flow changes the temperature, and this change is detected. This is shown in Figure 8.6. In MEMS devices a number of solutions have been found. The heater can be a resistor or a transistor and the temperature sensor can be a thermopile, a platinum resistor or a diode. Among the main applications are wind speed on the surface of the planet Mars [53], in respirators, spirometers or metabolic testing systems [54] for a measurement range of 2 to 400 l/min. In MEMS applications channels are small and the resistors are very thin and small to provide fast response and sensitivity. The practical realizations involve micromachining the substrate in order to place the resistor suspended and in direct contact with the flow as shown in Figure 8.6 or directly mounted in one of the walls of the channel.

Microanemometers based on the thermal principle are subject to several constraints. The thermal conductivity and the electrical conductivity of the thin film material are not the same as in bulk material and can be even considerably lower [55]. Furthermore, the heat transfer macroscopic correlations that are commonly used to relate the convection coefficient with the Reynolds number [56], and hence with the flow velocity, may not be accurate in small dimensions [57]. The heat convection coefficient h is related to the Reynolds number by

$$h = \frac{Nu\,k}{L_c} \tag{8.30}$$

where Nu is the non-dimensional Nusselt number, k is the fluid thermal conductivity and L_c is a characteristic length for comparison purposes between geometries. As an example, for a flow around a cylinder the characteristic length is the diameter. The Nusselt number may be correlated to the Reynolds, Re, and Prandtl, Pr, non-dimensional numbers depending on the geometry of the flow. For example,

$$Nu = 0.3 + \frac{0.62Re^{1/2}Pr^{1/3}}{\left(1 + \left(\frac{0.4}{Pr}\right)^{2/3}\right)^{1/4}} \tag{8.31}$$

is used for the flow around a cylinder for values of $Pr > 0.2$, where

$$Re = \frac{\rho UD}{\mu} \tag{8.32}$$

and

$$Pr = \frac{\mu C_p}{k}, \tag{8.33}$$

in which μ is the fluid viscosity, C_p the specific heat at constant pressure, D is the cylinder diameter and U is the fluid velocity. Equation (8.31) relates the Nusselt number to the fluid velocity. There are basically two ways to extract the value of the fluid flow. One is to feed the resistor with a constant power and then measure the change in hot point temperature as a function of the fluid velocity, and the other is to set a control loop to maintain the same overheat δT above the air temperature and to measure the value of the required power to keep this target overheat for every fluid velocity. The first method is called the constant power method, and the second the constant temperature method.

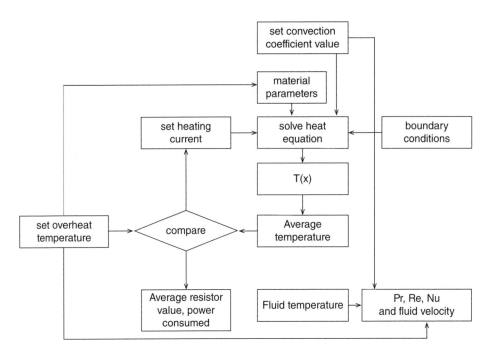

Figure 8.8 Computing power–flow velocity graphs

As an example, the constant power method can be implemented using a simple algorithm to calculate graphs of the fluid velocity as a function of the power consumed. The algorithm is shown in Figure 8.8. As can be seen, we set three values: (1) a target overheat temperature, $\Delta T = T_{HOT} - T_\infty$, where T_{HOT} is the average temperature of the heated resistor and T_∞ is the fluid temperature away from the boundary layer; (2) a heating current I; and (3) a convection coefficient value h. We select the heating resistor material parameters, resistivity and thermal conductivity, at the absolute temperature above the ambient. We solve the heat equation appropriate to the geometry of the problem with suitable boundary conditions. As a result, we end up with the temperature profile $T(x)$. We then calculate the average temperature $\overline{T}$ and compare $\overline{T} - T_\infty$ to $\Delta T = T_{HOT} - T_\infty$. If the calculated value of $\overline{T}$ is less than T_{HOT} then the current I is increased, and vice versa, until we reach a reasonable level of accuracy. From the value of $\overline{T} = T_{HOT}$ the value of the average resistance is calculated from the resistance–temperature equation of the heating resistor and the value of the electrical power dissipated in the resistor, I^2R. From the value of the convection coefficient h, we calculate the power lost to convection, $h\Delta TA_{conv}$. From the value of the fluid temperature T_∞ we calculate the values of the appropriate fluid parameters: thermal conductivity, specific heat at constant pressure, density and viscosity. Those magnitudes depend on temperature. From them we calculate the Prandtl number, and the Nusselt number from equation (8.30). We solve equation (8.31) to find the value of the Reynolds number, and finally, from equation (8.32), we get the fluid velocity U. This process flow is illustrated in Problems 8.3 and 8.4 and in Example 8.4.1.

8.4.1   Example: Micromachined Flow Sensor

In this example we follow the data and results reported in reference [58] where a microma-chined cantilever made of platinum is suspended in the middle of the flow in a channel. The platinum resistor is $L = 6\,\mu m$ long, $W = 800$ nm wide and $T = 40$ nm thick, resulting in a resistance value at 300 K of 305 Ω. The temperature coefficient of the resistor is $0.2\%/°C$. (1) Find the values of the two constants R_0 and A of a linear model for the temperature dependence of the resistance. (2) We know from measurements reported in [58] that at a flow velocity of $U = 1$ m/s the change in the voltage across the resistor when a 3 mA current is applied is 20 mV, and when the flow velocity is 3 m/s the change is 30 mV. Calculate the values of the average temperature of the resistor in each case. (3) Write the values of the convection and conduction sections A_{conv} and A_{cond}. (4) Write the heat equation. (5) Write Matlab code to solve the heat equation and to calculate the average value of the temperature distribution along the platinum resistor. (6) By a trial and error procedure, find the effective value of the heat convection coefficient that should be used in order to obtain the average temperatures for $U = 1$ m/s and $U = 3$ m/s calculated in (2) above.

(1) We assume that

$$R = R_0(1 + AT).\tag{8.34}$$

As we know the relative change in the resistance value when temperature changes, we can write

$$0.2 = \frac{1}{R}\frac{dR}{dT} \times 100 = \frac{1}{R}R_0A \times 100 = \frac{R_0A \times 100}{R_0(1+AT)} = \frac{A \times 100}{1+AT}.\tag{8.35}$$

We assume that this is valid at the reference temperature $T = 300$ K. Then $A = 0.0008\,\Omega/K$. With this value and knowing that at $T = 300$ K, $R = 305\,\Omega$, we get $R_0 = 245.9\,\Omega$.

(2) The voltage drop across the resistor when a constant current is applied is $V = IR$, so, at the reference temperature, $V = 3 \times 10^{-3} \times 305 = 0.915$ V. Hence at $U = 1$ m/s the voltage across the resistor will be $V(U = 1) = 0.915 + 0.02 = 0.935$ V and when $U = 3$ m/s,

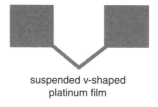

suspended v-shaped
platinum film

Figure 8.9   Example 8.4.1

$V = 0.915 + 0.03 = 0.945$ V. Then

$$R(U = 1) = \frac{0.935}{3 \times 10^{-3}} = 311.66\,\Omega, \tag{8.36}$$

$$R(U = 3) = \frac{0.945}{3 \times 10^{-3}} = 315\,\Omega. \tag{8.37}$$

The average temperature along the resistor is given by

$$\bar{T} = \left(\frac{R}{R_0} - 1\right)\frac{1}{A}. \tag{8.38}$$

For the two values of the flow velocity,

$$\bar{T}(U = 1) = \left(\frac{311.66}{245.9} - 1\right)\frac{1}{0.0008} = 333.8\,\text{K}, \tag{8.39}$$

$$\bar{T}(U = 3) = \left(\frac{315}{245.9} - 1\right)\frac{1}{0.0008} = 350.8\,\text{K}. \tag{8.40}$$

(3) The convection and conduction areas for a section of the resistor of length Δx are given by

$$A_{\text{cond}} = Wt \quad \text{and} \quad A_{\text{conv}} = 2\Delta x(W + t). \tag{8.41}$$

The resistance of this Δx can be written as

$$dR = dR_0(1 + AT) = \frac{R_0}{L}\Delta x. \tag{8.42}$$

(4) The heat equation is

$$\frac{d^2T}{dx^2} = \frac{2h(W + t)}{kWt}\Delta T - \frac{I^2 R_0}{kLWt}(1 + AT). \tag{8.43}$$

(5) The solution of equation (8.43) can be implemented in code as follows:

```
function dydx=flowfunctionbook(x,y)
dydx=[y(2)
2*h*(t+w)/(k*w*t)*(y(1)-Tair)-I^2*Ro*(1+A*y(1))/(k*L*w*t)];
```

with the boundary conditions of the temperature at both ends in equilibrium with the medium.

(6) The average temperature is evaluated once the temperature distribution is calculated by means of

```
meantemp= trapz(x,y(1,:))/L
```

By a trial and error procedure we find that for $U = 1$ m/s the effective value required for h is $h = 7 \times 10^6$, while for $U = 3$ m/s it is $h = 4.5 \times 10^6$.

8.5 Dynamic Thermal Equivalent Circuit

The heat equation described in Section 8.2 corresponds to the steady state. The study of simplified relevant dynamic response is possible with a lumped-element equivalent circuit extending the one described in Figure 8.3 and including time-dependent components, in particular equivalent thermal capacitors. Returning to equation (8.44) and considering the same volume shown in Figure 8.2 in non-steady-state conditions, the change in internal energy U_{int} of the volume in a differential of time can be written as

$$\frac{dU_{int}}{dt} = \dot{Q}_{in} + \dot{Q}_{gen} - \dot{Q}_{out} - \dot{Q}_{conv}. \tag{8.44}$$

Following the same steps as in Section 8.2,

$$\frac{dU_{int}}{dt} = k\frac{d^2}{dx^2}\Delta x A_{cond} + \dot{Q}_{gen} - \dot{Q}_{conv}. \tag{8.45}$$

The change in internal energy of the body is

$$\frac{dU_{int}}{dt} = c\,\rho_m vol\frac{dT}{dt} \tag{8.46}$$

where ρ_m is the material density, vol is the volume and c is the specific heat. Equation (8.45) becomes

$$c\rho_m vol\frac{dT}{dt} = k\frac{d^2}{dx^2}\Delta x A_{cond} + \dot{Q}_{gen} - \dot{Q}_{conv}. \tag{8.47}$$

Under certain conditions the body can be assumed to be isothermal, and then

$$c\,\rho_m vol\frac{dT}{dt} = \dot{Q}_{gen} - \dot{Q}_{conv}. \tag{8.48}$$

Taking into account the convection component,

$$c\,\rho_m vol\frac{dT}{dt} = \dot{Q}_{gen} - hA_{conv}(T - T_\infty). \tag{8.49}$$

Equation (8.49) can be compared to the equation that can be written in the equivalent electric circuit shown in Figure 8.10, where we apply the same equivalences as in Figure 8.3, which means that $\dot{Q}$ is equivalent to the electric current I, the voltage is equivalent to the electric potential, C_{th} is equivalent to an electrical capacitance and the thermal resistance Θ is equivalent to the electrical resistance:

$$C_{th}\frac{dT}{dt} = \dot{Q}_{gen} - \frac{T - T_\infty}{\Theta_{conv}}. \tag{8.50}$$

Figure 8.10   Equivalent dynamic circuit

The equations giving the values of the thermal capacitance and thermal resistance due to convection are

$$C_{\text{th}} = c\,\rho_m vol \quad \text{and} \quad \Theta = \frac{1}{hA_{\text{conv}}}. \tag{8.51}$$

8.6   Thermally Actuated Bimorph

Stacks of two different materials are know as bimorphs and have applications in thermally sensitive actuators in the micro and macroworld. In the microworld thin layers of stacked materials that have different temperature expansion coefficients are commonly used. In such structures the cross-section of a beam or cantilever is composed of two different materials having different Young's moduli and other parameters. This problem was first studied by S. Timoshenko, and following the derivation in [59] one can assume that forces F_1 and F_2 arise at the section shown in Figure 8.11, and it can be assumed that they are applied at the middle of the respective material thickness. Given the dimensions shown in Figure 8.11, the equilibrium of forces and moments in the section leads to

$$F_1 = F_2 = F, \quad M_{b1} + M_{b2} = F_1\frac{a_1}{2} + F_2\frac{a_2}{2} = \frac{Fh}{2}. \tag{8.52}$$

The bending moments M_{b1} and M_{b2} are related to the radius of curvature ρ_c. If each of the materials has Young's modulus E_1 and E_2, respectively, then

$$M_{b1} = \frac{E_1 I_1}{\rho_c}, \quad M_{b2} = \frac{E_2 I_2}{\rho_c}, \tag{8.53}$$

which together with equation (8.54) gives

$$\frac{E_1 I_1 + E_2 I_2}{\rho_c} = \frac{Fh}{2}. \tag{8.54}$$

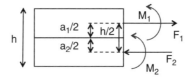

Figure 8.11   Bimorph. Forces and moments

The unit elongation of the two materials at the common surface must be equal, thus the total strain of the two materials must be equal as the initial length is the same. We have three components of the total strain $\epsilon = \epsilon_{11} + \epsilon_{12} + \epsilon_{13}$, and we calculate them for the upper part of the bimorph:

1. The elongation due to the thermal expansion is

$$\epsilon_{11} = \alpha_1 \Delta T, \tag{8.55}$$

   where α is the thermal expansion coefficient.
2. The force F_1 is assumed to be applied at the centre of the thickness of the upper part of the beam. The strain due to this force, and thus the elongation due to the section forces, is

$$\epsilon_{12} = \frac{F_1}{W a_1 E_1}, \tag{8.56}$$

   where W is the width of the beam (not shown in the figure). As we are interested in the strain at the common surface of the two materials of the bimorph we have to add the third component that is calculated according to Figure 8.12.
3. We see from Figure 8.12 that

$$\tan \theta = \frac{\Delta L}{a_1/2}, \tag{8.57}$$

and the arc and the angle are related by

$$L = \theta \rho_c. \tag{8.58}$$

For small deformations $\tan \theta \simeq \theta$, we have

$$\tan \theta = \frac{\Delta L}{a_1/2} = \frac{L}{\rho_c}, \tag{8.59}$$

and then

$$\epsilon_{13} = \frac{\Delta L}{L} = \frac{a_1}{2\rho_c}. \tag{8.60}$$

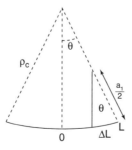

Figure 8.12 Third component of the strain

Setting the two total strains equal,

$$\alpha_1 \Delta T + \frac{F_1}{W a_1 E_1} + \frac{a_1}{2\rho_c} = \alpha_2 \Delta T - \frac{F_2}{W a_2 E_2} - \frac{a_2}{2\rho_c}, \tag{8.61}$$

where we have assumed that the widths of the two parts of the bimorph are the same. Using equations (8.54) and (8.61),

$$\frac{1}{\rho_c} = \frac{(\alpha_2 - \alpha_1)\Delta T}{\frac{h}{2} + \frac{2(E_1 I_1 + E_2 I_2)}{h}\left(\frac{1}{W_1 a_1 E_1} + \frac{1}{W_2 a_2 E_2}\right)}. \tag{8.62}$$

Taking into account that

$$I_1 = \frac{W a_1^3}{12}, \quad I_2 = \frac{W a_2^3}{12}, \tag{8.63}$$

we obtain

$$\frac{1}{\rho_c} = \frac{6(\alpha_1 - \alpha_2)\Delta T(1 + m)^2}{h(3(1 + m)^2 + (1 + mn)(m^2 + \frac{1}{mn}))}. \tag{8.64}$$

In equation (8.64), $m = a_1/a_2$ and $n = E_1/E_2$. Once the radius of curvature is known, tip deviations, v_{end}, can be calculated from trigonometric relationships,

$$\rho_c \cos\theta = \rho_c - v_{end}, \quad L = \theta\rho_c, \quad v_{end} = \rho_c\left(1 - \cos\frac{L}{\rho_c}\right). \tag{8.65}$$

A leverage effect can be achieved by using long values for L.

8.6.1   Example: Bimorph Actuator

In [60] a thermal bimorph micromirror with large bidirectional and vertical actuation is described. The technology process is outlined in Section 9.9 below. Find values for the thermal expansion coefficients and Young's modulus of aluminium and polysilicon, and calculate the radius of curvature as a function of ΔT. Assume that the bimorph length is 200 μm and that the two thicknesses are equal, $a_1 = a_2$.

The Young's modulus of polysilicon is $E_2 = 169$ GPa (Table 2.5) and of aluminium is $E_1 = 69$ GPa. For the values given in the problem, $m = 1$ and $n = 0.408$. The thermal expansion coefficient for aluminium is $22.2 \times 10^{-6}\,\mathrm{m/m \cdot K}$, and we have taken that of single crystal of silicon for the polysilicon film $\alpha_2 = 2.6 \times 10^{-6}\,\mathrm{m/m \cdot K}$. Using equation (8.64), we find that

$$\rho_c = \frac{0.068}{\Delta T}\,\mathrm{m}, \tag{8.66}$$

and the end deflection of the tip is given by equation (8.65). For $\Delta T = 10°C$, we obtain $v_{end} = 3.4\,\mu m$.

8.7   Thermocouples and Thermopiles

A thermocouple is a device consisting of two wires of different materials A and B brought together as shown in Figure 8.13. When a couple of wires of, in general, a third material C are connected to the free ends of A and B and these connections are held at a reference temperature that is different from the temperature at the point where A and B are bonded, a voltage is generated that is proportional to the temperature difference ΔT. This effect is known as the Seebeck effect, and the voltage generated is written

$$V = S_{AB}\Delta T, \tag{8.67}$$

where S_{AB} is the relative Seebeck coefficient between the two materials (the difference in the two absolute values) and $\Delta T = T - T_{ref}$ is the temperature difference. As the voltage generated by a single thermocouple is small, this same idea is extended to an array of N thermocouples arranged in series, called a thermopile.

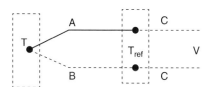

Figure 8.13   Thermocouple: material A (thick solid line), material B(thin solid) and material C (dashed)

8.7.1   Example: IR Detector

In [61] a thermopile is described made out of $N = 240$ series Ti/Ni thermocouples fabricated by evaporation of thin films of the two metals on top of a membrane fabricated by bulk micromachining of silicon (see Chapter 9). The thermocouples sit on top of a silicon dioxide layer as electrical insulator and are placed in such a way that one side of the thermocouple junctions is located at the IR radiation exposed area on top of the thin membrane, while the

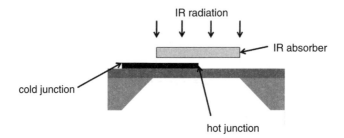

Figure 8.14   IR detector with thermopile in a membrane

other, or reference junctions, are located on top of the thick frame of full thickness of the silicon wafer. The hot area is covered by a IR radiation absorber film as shown in Figure 8.14. The output voltage at a given radiation of 1 *W is* 59 *mV. Estimate the value of* ΔT *in the membrane.*

The output voltage of the series association of the 240 thermocouples is

$$V = NS_{AB}\Delta T. \tag{8.68}$$

The Seekeck coefficient difference between titanium and nickel is $S_{AB} = 14.8 \times 10^{-6}$ V/K. Therefore

$$V = 240 \times 14.8 \times 10^{-6} \times \Delta T = 59\,\text{mV}, \tag{8.69}$$

and so ΔT= 16.6 K. The temperature difference between the hot and cold junctions is provided by the membrane, The hot junctions are in the membrane and the cold junctions are in the frame. Due to the fact that the IR power heats the membrane and that the membrane is thin compared to the frame, it can be assumed that the frame is at room temperature and that there is a temperature distribution inside the membrane with boundary condition at the edges similar to the problem described in Section 8.3.1, in such a way that the average temperature increase is 16.6 K.

Problems

8.1   We have a die of single-crystal silicon with a platinum resistor patterned on its surface, as shown in Figure 8.15. The resistor has a value of $R(350\,\text{K}) = 100\,\Omega$, thickness $t_R = 50\,nm$ and width of $W_R = 5\,\mu\text{m}$. We also know that the resistivity of the platinum film at 350 K is $\rho = 1.281 \times 10^{-7}\,\Omega\,\text{m}$. Assuming that the heat is generated from the dissipation of power when the resistor is fed with a constant current I, and that it is evenly distributed across the silicon die, write the equation of the resistance of the Δx shown in Figure 8.15.

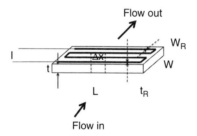

Figure 8.15   Problem 8.1

8.2   Write the heat equation for the geometry shown in Figure 8.15, with dR as found in Problem 8.1. Assume that the resistor takes a value of $R = 100\,\Omega$ at a temperature $T_{HOT} = 350$ K.

8.3   For the heat equation found in Problem 8.2, find suitable boundary conditions if the two ends of the silicon die are attached by means of two supports of the same cross-section but with different thermal conductivity, as shown in Figure 8.16.

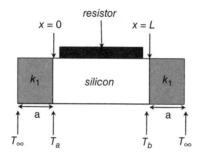

Figure 8.16   Problem 8.3

8.4   For the same geometry as in Problem 8.3, write Matlab code to solve the differential equation with the boundary conditions of Problem 8.3, assuming that the silicon die has length $L = 1000\,\mu$, thickness $t = 100\,\mu$m and width $W = 1000\,\mu$m, and for heat convection coefficient $h = 100, 200, 500$ W/m^{2}C.

8.5   For the same geometry as in Problem 8.4, we would like to assess the effect of the value of the thermal conductivity of the supports, k_s, on the power balance. Check using 0.1 and 1 W/m · K as values for k_s.

8.6   We have a heated volume of silicon with a power of 1 mW using a resistor. Imagine that the silicon volume is assumed to be isothermal. The volume is 1000 μm square by 500 μm thick. This volume is anchored by a thinner and narrower silicon bridge 1000 μm long, 5 μm wide and 2 μm thin. Assume that the convection heat transfer coefficient is $h = 100$ and that all the surface of the silicon volume is subject to the convective heat transfer. Calculate the silicon volume temperature if the air temperature is $T_\infty = 300$ K, and the frame temperature where the bridge is anchored is 310 K. Calculate the amount of power required to heat the silicon temperature up to 340 K. The thermal conductivity of silicon is assumed to be 133.5 W/m·K.

8.7 In many thermal based MEMS devices, thermal isolation is needed around one area that should be at a higher temperature than its surroundings [62]. This can be achieved by interposing thermally isolating supports as illustrated in Problem 8.3, or by creating a bridge thinner than the supporting frame. This is illustrated in Figure 8.17. Write the steady-state heat equation.

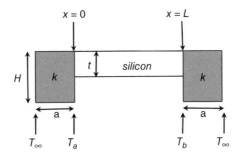

Figure 8.17   Problem 8.7

8.8 For the same geometry as in Problem 8.7, assume that the power given to the device is $P = 10$ mW and that the thermal conductivity of the bridge and frame are the same $k = 133.5$ W/m·K. The bridge has thickness $t = 2\,\mu$m, length $L = 100\,\mu$m and width $W = 10\,\mu$m, whereas the frame has thickness $H = 500\,\mu$m, length $a = 1000\,\mu$m and width $W = 10\,\mu$m equal than that of the bridge. Write a Matlab code to solve for the temperature distribution along the bridge, and plot the result.

8.9 Consider the geometry shown in Figure 8.18, where there is a rod made of a material assumed to be a thermal insulator and covered with a thin film of a resistive material. Calculate the equations of the conduction and convection areas for a segment of the rod of length Δx. Write the equation of the differential resistance, dR, for this same segment. Write the steady-state heat equation.

8.10 Write a Matlab code to solve the steady-state heat flow equation derived in Problem 8.9 with a boundary condition at the position $x = L$, $dT/dx = 0$. Solve this boundary value

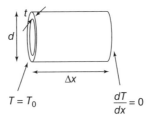

Figure 8.18   Problem 8.9

problem for the following parameter values: $t = 0.635 \times 10^{-6}$, $k = 75$ W/m·K, $L = 1.02 \times 10^{-2}$ m, $I = 30$ mA and resistivity of the thin film material $\rho = 10.6 \times 10^{-8}\ \Omega$ m.

8.11 We have a flow meter in which a silicon volume of 1 mm^2 and 500 μm thick is heated by a platinum resistance. The chip is supported by two arms of length 1 mm, thermal conductivity 1.005 W/m·K and cross-section $A = 50 \times 50$ μm^2. A flow is blown on the mass and the convection coefficient is $h = 200$. Write PSpice code for the thermal circuit and solve the transient response, assuming that the power injected is 10 mW and that the fluid temperature is 300 K. The two supporting arms are connected to a PCB the temperature of which is assumed to be in equilibrium with the air, $T_{PCB} = 300$ K. The specific heat is $c = 1.005$ W/m·K.

9

Fabrication

This chapter is brief summary of the main fabrication process steps with brief descriptions. The number of processing technologies and the combination of process steps is very large and depends very much on the facilities used and on the experience of the researchers in given procedures. It is beyond the scope of this book to go into great detail. However, a full understanding of the potential of MEMS technology will be within easier reach after reading an outline of the fundamental process steps, illustrated with examples and problems. This is the objective of this chapter. Fundamental process steps are photolithography and the growth, deposition and etching of materials. Relevant questions are the anisotropy and selectivity of the etching processes.

9.1 Introduction

MEMS technology involves many processes that are common to the microelectronics industry, while incorporates specific process steps to provide the capability to create membranes or release moving parts that are not present in mainstream CMOS technology. The technology operations involved in a MEMS process are as follows:

1. Material growth. Thermal oxidation and epitaxial growth are the most frequently used material growth techniques. Thermal oxidation of silicon is a very common process as good-quality oxide can be grown on the wafer surface, though at high temperatures. Silicon dioxide is a dielectric, and besides being used to create MOS transistor gates, also serves as an insulating layer between devices or between a device and the substrate. Epitaxial growth is used to provide silicon on insulator wafers instead of a wafer bonding, grinding and polishing process [63].
2. Material deposition. Many materials need to be deposited on top of a substrate or other layers. They can be dielectric (silicon dioxide (SiO_2), silicon nitride (Si_3N_4), hafnium oxide (HfO_2), conductors (gold, aluminium, chromium, titanium, silver, platinum, . . .) semiconductors (polysilicon, amorphous silicon) or polymers (resist, SU-8, PDMS, parylene, . . .), and they are deposited using a number of technologies such as thermal evaporation,

Understanding MEMS: Principles and Applications, First Edition. Luis Castañer.
© 2016 John Wiley & Sons, Ltd. Published 2016 by John Wiley & Sons, Ltd.
Companion Website: www.wiley.com/go/castaner/understandingmems

sputtering, low-pressure chemical vapour deposition (LPCVD), plasma-enhanced chemical vapour deposition, spin-on, spray, electrospray).

3. Etching. Etching of materials can be isotropic or anisotropic and the methods used are wet or dry. Selectivity of the etchants is a very important property that allows proper design of particular process flow. Dry processing allows a larger aspect ratio (amount etched in depth compared to amount etched horizontally)[1] for trenches and walls.

4. Patterning. Patterning materials, either sacrificial or structural, is an essential part of a MEMS process. The sequential use of a set of masks allows the proper manufacture of a MEMS device.

9.2 Photolithography

One of the fundamental processes in microelectronic manufacturing is photolithography. This is schematically shown in Figure 9.1. A starting substrate (S) is covered by a thin layer of photoresist by spin-on coating and cured at a moderate temperature. The photoresist-covered substrate is then subject to UV light exposure through a mask (MK) which is a transparent glass with a dark coating in specific areas. UV light sensitizes the photoresist in those areas not covered by the coating, the rest remaining unsensitized. This step is performed in a mask aligner. A developer is used in the following step to dissolve the sensitized photoresist, leaving the rest intact. At the end of this process the substrate is protected by the photoresist layer in specific areas. Several masks are usually required to manufacture a device, and the mask aligner is used several times. As the masks are normally different from each other, alignment aids are also included in each mask to make sure that the sequential operations are made in the right places.

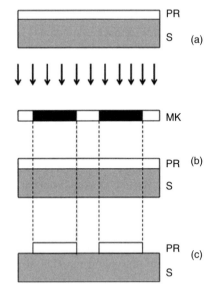

Figure 9.1 Photolithography process for a positive photoresist, PR = photoresist, S = substrate, MK = mask. (a) Photoresist deposition, (b) photoresist exposure and (c) photoresist developing

---

[1] This definition of the aspect ratio used here for the anisotropic etching differs from that in Chapter 4 used for the layout of a resistor.

The process shown in Figure 9.1 is for a positive photoresist, which is a type of resist that becomes soluble in the developer after exposure to UV light. In contrast, if a negative photoresist is used, the result is exactly the opposite to that shown in Figure 9.1, as a negative photoresist becomes insoluble in the developer after being exposed to the UV light. Exposure can also be done using shorter wavelengths or electron beams. One of the negative photoresists is known as SU-8 and is very popular in MEMS processing because it has the property of being very hard to strip and is used where an unstrippable layer is required.

9.3 Patterning

The photolithography process is normally used to perform operations in those areas uncovered by the resist. One typical example is to pattern a thin layer of material as shown in Figure 9.2.

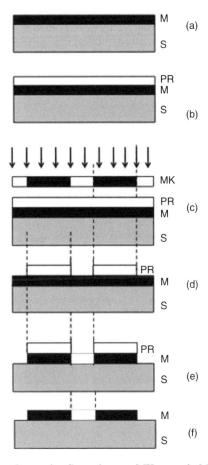

Figure 9.2 Patterning, PR = photoresist, S = substrate, MK = mask, M = material. (a) The material M is deposited on top of the substrate. (b) The photoresist is spun on top of the material M and cured. (c) The photoresist is exposed through a mask. (d) The exposed photoresist is stripped with a developer. (e) The exposed material areas are etched through the windows in the photoresist. (f) Any unexposed photoresist is dissolved

As can be seen, the material M is first deposited on top of a substrate, and then covered by the photoresist layer that undergoes the light exposure step as described earlier. If the photoresist is positive, the dissolved photoresist areas cover the parts of the material M shown in Figure 9.2(d). The windows opened in the resist allow etching of the material M (in step (e)) followed by stripping of the remaining resist (in step (f)). Care should be taken that the etching procedure is selective and does not also etch the resist or substrate. Selectivity is a very important property of the etchants for the microelectronics and MEMS processing.

9.4   Lift-off

One special case of material patterning is known as the lift-off technique. This is illustrated in Figure 9.3. As can be seen, the photoresist is patterned after exposure through the mask. The material to be patterned is then deposited on top of the already patterned resist. Care should be taken to use a thick enough layer of photoresist such that the thickness of the material deposited is less than the thickness of the photoresist. If such is the case, the side faces of the photoresist are exposed at the edges and hence when the resist is dissolved, the material that was laid on top will also be removed.

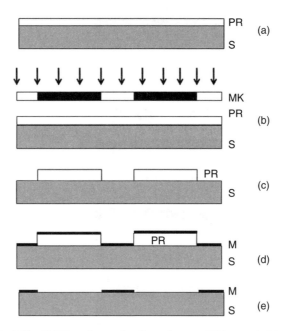

Figure 9.3   Lift-off. PR = photoresist, S = substrate, MK = mask, M = material

9.5   Bulk Micromachining

Bulk micromachining is the technological operation carried out to remove a significant amount of bulk material in specific areas of a substrate. Normally a hard mask is required to prevent unwanted etching outside the designated areas. Sometimes the same photoresist can act as hard mask in some of the etching processes, but also a non-photosensitive material may be

Figure 9.4   Bulk micromachining vertical walls. S = substrate, HM = hard mask

required for specific etchants. If we pattern a hard mask material following the same steps summarized in Figure 9.2, we can then proceed to etch the bulk substrate material. Some etching processes are isotropic and then the removal of bulk material occurs in all directions even under the hard mask. However, in most cases, the etching process is anisotropic and hence has preferential directions. Among the etching processes that can be found, there are wet and dry etching procedures. Wet etching uses suitable reagents while dry etching uses reactive ion etching equipment with a specific sequence of plasma etchants. Dry processes end up with almost vertical walls as shown in Figure 9.4. In wet anisotropic etching processes of bulk crystalline materials such as silicon, the etch rate depends on the crystallographic plane orientation, being much larger in specific directions. This ends up in tilted walls where the low etching crystallographic plane forms at the surface, as can be seen in Figure 9.5.

Figure 9.5   Bulk micromachining vertical walls S = substrate, HM = hard mask

The two properties of any etching process, selectivity and anisotropy, are crucial to designing a MEMS process flow. For example, silicon bulk material can be etched by both wet or dry etching steps. Among the wet silicon etchants potassium hydroxide (KOH) and tehtramethyl ammonium hydroxide (TMAH) are the most commonly used. The two are anisotropic and do not leave vertical walls because the etch rate for planes (100) is much larger than that for planes (111) for the two wet etchants.

9.5.1   Example: Angle of Walls in Silicon (100) Etching

For a wet etching process that uses a KOH solution having 50 μ m/h etch rate, calculate the angle that forms the walls and the time required to etch 200 μm deep from the surface of a single crystal silicon wafer (100). We know that the etch rate ratio of (100) to (111) planes is 30. Calculate the side etching in the direction of the (111) plane at the end.

The etch rate of the (111) plane is much slower than that of the (100) plane, so the (100) face will be etched much faster that the (111), hence the inclination will be the crystallographic angle of the two planes. We have already calculated this angle in Example 2.6.1, and it is 54.73°. The time required to etch 200 μm is 4 hours and the etch in the direction of the normal to the (111) plane is $50/30 \times 4 = 6.66$ μm.

9.6 Silicon Etch Stop When Using Alkaline Solutions

In the fabrication of membranes or other micromechanical structures, several mechanisms can be implemented to stop the etch when specified to have a given thickness for the structural layer. These techniques involve: timed, p^+ doping, photovoltaic, galvanic electrochemically controlled pn junctions among others [64, p. 39]. We briefly summarize here the p^+ etch stop. Adding III group impurities to silicon greatly reduces the etch rate. For example, for silicon (100) in a 10% KOH solution at 60°C, the etch rate is reduced by almost two orders of magnitude if the boron concentration exceeds $1 \times 10^{19} \text{cm}^{-3}$. One way to implement an etch stop technique involves creating a thin layer of boron-rich silicon by ion implantation and drive-in at high temperature. If the total amount of impurities implanted is S (impurities per square centimetre), the drive-in step redistributes them, creating a Gaussian impurity concentration distribution

$$C(x) = \frac{S}{\sqrt{\pi Dt}} e^{-x^2/4Dt}, \qquad (9.1)$$

where D is the diffusivity of the impurities in silicon at the drive-in temperature (in square centimetres per second), and t is the drive-in time.

9.6.1 Example: Boron drive-in at 1050°C

Boron impurities totalling $S = 1 \times 10^{21}$ cm^{-2} have been implanted into bulk silicon (100). Calculate the time required to drive in these impurities such that the concentration at 0.5 μm depth is 1×10^{19} cm^{-3}.

The diffusivity of boron in silicon at 1050°C is $D = 9.3 \times 10^{-14}$ cm^2/s.[2] Then, using equation (9.1), we can write

$$1 \times 10^{19} = \frac{1 \times 10^{21}}{\sqrt{\pi \times 9.3 \times 10^{-14} t}} e^{(-0.5 \times 10^{-4})^2/4 \times 9.3 \times 10^{-14} t}. \qquad (9.2)$$

The resulting drive-in time is around 7 minutes.

9.7 Surface Micromachining

Many MEMS devices are manufactured today using a surface micromachining process as shown in Figure 9.6. A sacrificial layer is deposited on top of the substrate and patterned to provide anchor space. A structural layer is deposited on top and also patterned. Finally, the sacrificial layer is removed, leaving, in the example shown in Figure 9.6, a bridge of structural layer

---

[2] http://fabweb.ece.illinois.edu/gt/gt/gt10.aspx

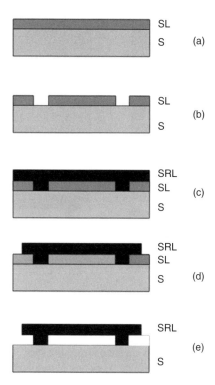

Figure 9.6 Surface micromachining, S = substrate, SL = sacrificial layer, SRL = structural layer. (a) Sacrificial layer deposition, (b) sacrificial layer patterning, (c) structural layer deposition, (d) structural layer patterning and (e) sacrificial layer etch

anchored to the substrate. In this process, selectivity of the the etchants between substrate sacrificial layer and structural layer is required. Notice that there is no substrate material removal.

9.7.1 Example: Cantilever Fabrication by Surface Micromachining

A cantilever can be fabricated in a variety of ways. In this example we adapt the fabrication process described in [65]. The process uses amorphous silicon doped with phosphorus as structural layer, silicon dioxide as sacrificial layer, chromium as electrode material, silicon nitride as isolation layer on top of the substrate and hafnium oxide as high-k dielectric. The process is outlined in Figure 9.7. We need two electrodes to be able to electrostatically bend the cantilever, and in this case the two are located on top of the isolation layer, which is silicon nitride (a dielectric). In order to prevent any short-circuit between the two electrodes a high-k dielectric such as hafnium oxide is deposited. The structural layer is anchored at one point in the figure, and this anchor also serves as the electric contact to the upper plate of the cantilever actuation.

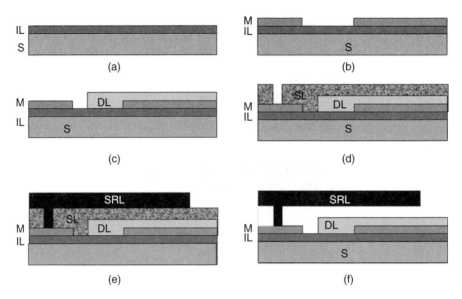

Figure 9.7 Surface micromachining, S = substrate, IL = insulating layer, M = metal, DL = dielectric layer, SL = sacrificial layer, SRL = structural layer. (a) Insulating layer deposition (silicon nitride), (b) metal (chromium) layer deposition and patterning, (c) dielectric layer (hafnium oxide) deposition and patterning, (d) sacrificial layer (silicon dioxide) deposition and patterning, (e) structural layer (amorphous silicon) deposition and patterning and (f) release of cantilever by sacrificial layer etch

9.8   Dry Etching

Dry etching is used to anisotropically etch materials and uses plasma chemistry. This is known as deep reactive ion etching (DRIE). It achieves high aspect ratios. For silicon a fluorine-based chemistry is used, whereas for GaAs a chlorine-based chemistry is used instead. One very popular MEMS process is the Bosch [66] process consisting of two steps that are sequentially repeated: one etch step using SF_6 gas followed by a deposition step based on octofluorocyclobutane (C_4F_8). The deposition step yields an inert passivation layer similar to Teflon. The Bosch process can achieve a 30 : 1 aspect ratio, and photoresist selectivity against etching up to 75 : 1 is claimed, while silicon oxide selectivity can be as large as 200 : 1. This is why resists, silicon dioxide and even alumina can be used to mask the etching process.

9.9   CMOS-compatible MEMS Processing

A number of different approaches to incorporating MEMS devices into mainstream CMOS technology have been tried. Most involve post-processing of the CMOS wafers once the circuit has been fabricated. The post-processing involves protection of the CMOS circuit area against silicon and oxide etches, and the metal and polysilicon layers that have been previously patterned act as structural and hard mask layers. Of course each CMOS process has its specificity and hence the post-processing depends on the built-in CMOS layer properties.

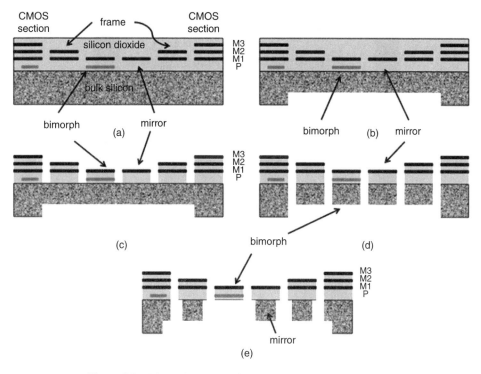

Figure 9.8   Bimorph actuatior for a mirror, adapted from [60, 67]

9.9.1   Example: Bimorph Actuator Compatible with CMOS Process

In this example we adapt the information given in reference [60]. The CMOS process is an Agilent 0.5 µm with three layers of metal (M1, M2 and M3 in Figure 9.8) and one layer of polysilicon (P). This technology is available at MOSIS.[3] All etching process steps are dry.

The starting wafer has a CMOS section depicted in two edges of Figure 9.8(a) where the three metal layers and the polysilicon layer are shown embedded in silicon dioxide. The first post-processing step involves a membrane fabrication in the designated area where the bimorph actuator and mirror are to be placed (see Figure 9.9(b)). The silicon dioxide is a mask against the dry etching of the back. The second step (Figure 9.8(c)) is a dry etching of the silicon dioxide from the top. This is also a dry etch with high directivity, and the M3 layer is used as hard mask against this etch to protect designated areas for frame, bimorph and mirror besides the CMOS sections. In the third step (Figure 9.8(d)) a dry deep etch of silicon is performed and M3 and M2 act as hard mask at the designated areas. Finally, an isotropic silicon etch is performed that undercuts the silicon below the remaining oxide layer, and in particular below the bimorph the silicon is totally removed due to the small width of the fingers shown in Figure 9.9.

---

[3] https://www.mosis.com/

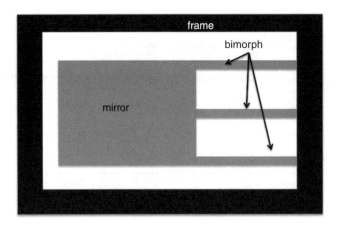

Figure 9.9   Bimorph actuator for a mirror, adapted from [60, 67]

9.10   Wafer Bonding

Silicon direct bonding is of great interest for creating silicon on insulator (SOI) wafers, and the basic process involves putting together the surface of two wafers and annealing at high temperature in an oxidation atmosphere. This results in stress-free multi-wafer structures that behave as monolithic [68]. Direct bonding can be carried out between two bare silicon wafers, between one bare wafer and an oxidized wafer or between two oxidized wafers. For integrated circuit applications, having a buried oxide between the two silicon wafers is very useful. The bond and etch back SOI (BESOI) process involves the direct bonding of two wafers with a silicon dioxide layer in between. If one if the two wafers is thinned down to few micrometres, we end up with a handle wafer, an oxide box a few micrometres thick on top, and a device layer of single crystal silicon. As the silicon etchants etch the silicon dioxide significantly less, the buried oxide acts as an etch stop, and hence membrane or beam fabrication is possible with automatic etch stop. This process is illustrated in Figure 9.10. As can be seen in Figure 9.10, the oxidized wafer in the BESOI process is a handle wafer. whereas the upper wafer is thinned down using a chemical-mechanical polishing (process that allows the preparation and finishing of very smooth surfaces). A competing technique is the separation by ion implant of oxygen (SIMOX) process, shown in Figure 9.10(b), where only one wafer is involved and a deep implantation of oxygen is done before an annealing. This creates a buried oxide layer a few micrometres deep from the surface.

Another technique for bonding wafers is anodic bonding, which requires the application of a voltage. It is useful for bonding an electron conductive material, such as silicon, to an ion-conductive material such as glass.

9.11   PolyMUMPs Foundry Process

MUMPs is a multi-project wafer commercial process offering several different full processes for different applications. PolyMUMPs is one the the processes offered, including three

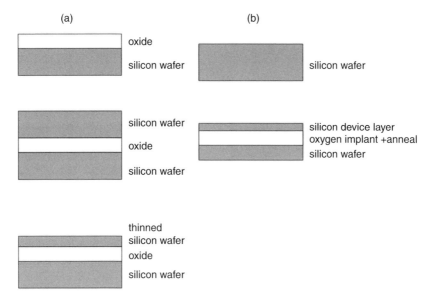

Figure 9.10 SOI process (a) BESOI and (b) SIMOX

polysilicon layers as structural layers, two oxide sacrificial layers and one metal layer.[4] Eight mask levels are available to create up to seven physical layers as described in [69]. This is a surface micromachining process that has many applications. In the following example the key steps of the process are shown.

9.11.1   Example: PolyMUMPs Cantilever for a Fabry–Perot Pressure Sensor

The measurement principle is to detect the reflection of a laser beam in a cantilever that bends. The cantilever is driven at close to resonance frequency. The air viscosity is a function of the ambient pressure and the damping coefficient is accordingly sensitive to pressure. A laser is used as a way of measurement specially adapted to harsh environments. Full measurement principles are described in [70].

We have adapted the information provided in [70] to summarize the PolyMUMPs technology processing. Only the nitride (N), level 1 polysilicon (P1) and two sacrificial layers of oxide (OX1 and OX2) are used. Some of the main process steps are shown in Figure 9.11.

The silicon substrate is an n-type (100) wafer which is doped with phosphorus. The following step is the deposition of 600 nm of silicon nitride (N) followed by a level 0 polysilicon layer (P0). As the P0 layer is not used in this device it is completely removed in the next mask step.

---

[4] http://www.memscap.com/products/mumps/polymumps

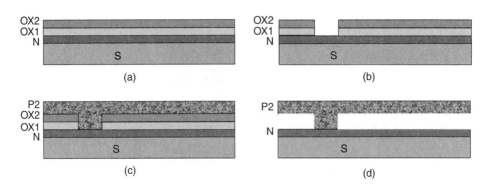

Figure 9.11 PolyMUMPs process, adapted from [70]

The next step is to deposit the first oxide layer (OX1) followed by the deposition of the level 1 polysilicon (P1). This level of polysilicon is not used in this process and hence is completely removed in the photolithography step. Next the second level of oxide (OX2) is deposited; it sits directly on top of the first oxide as the poly 0 and poly 1 layers have been removed. This is shown in Figure 9.11(a). The stack of the two oxides is then patterned. There are several masks in the process that can be used for this purpose, and in the example described here the anchor mask level is selected. This is shown in Figure 9.11(b). The oxide etch stops automatically when the silicon nitride layer is reached due to the selectivity of the etchant. After the window has been opened in the oxide stack, the level 2 polysilicon layer is deposited as shown in Figure 9.11(c) and patterned. Finally, the cantilever is released by etching the remaining areas of oxide stack as shown in Figure 9.11(d).

Problems

9.1 Find an analytical model for dry oxidation of silicon and find the thickness of oxide that will grow at 1000°C in dry atmosphere in 1 hour.

9.2 We want to protect an area of bulk silicon oriented (100) against KOH. Calculate the thickness of the oxide necessary to prevent etching the protected area while 200 μm of the unprotected area are etched. The concentration of KOH solution is 20% and the temperature 90°C.

9.3 In Example 9.7.1 an isotropic etch is performed to totally undercut the silicon below the bimorph area. Calculate the minimum time required if a hydrofluoric–nitric–acetic acid mixture etching solution is used.

9.4 In Example 9.6.1, for the fabrication of a cantilever, silicon dioxide is used as a sacrificial layer. The release of the cantilever requires the silicon dioxide to be etched selectively. From Figure 9.7, identify the materials that are in contact with the silicon dioxide before the release step and assess the selectivity of the silicon dioxide etch.

9.5 Calculate the concentration of impurities at a depth of 0.5 μm for drive-in times of 15 minutes if the total amount of impurities implanted is 5×10^{18} cm^{-2}. The drive-in temperature is 1050°C and the diffusivity is $D = 9.3 \times 10^{-14}$ cm^2/s.

9.6 For a diffusion of impurities into a bulk material from an unlimited source, the depth follows a complementary error function distribution. Calculate the total amount of boron impurities diffused into silicon after 30 minutes diffusion at 900°C.

9.7 For the same data as in Problem 9.6, calculate the junction depth if such diffusion is carried out on silicon doped with $N_B = 10^{17}$ cm^{-3} n-type.

9.8 Using the same data as in Problem 9.6, calculate the value of the sheet resistance resulting after the diffusion process described in Problem 9.6. Assume that the mobility of holes is given by

$$\mu = \mu_{min} + \frac{\mu_{max} - \mu_{min}}{1 + (N(x)/N_r)^{\alpha}}. \tag{9.3}$$

For boron diffusion into silicon, the values of the constants in equation (9.3) are $\mu_{min} = 44.9$ cm^2/Vs, $\mu_{max} = 470.5$ cm^2/Vs, $N_r = 2.23 \times 10^{17}$ cm^{-3} and $\alpha = 0.719$.

9.9 In [71] a membrane strain sensor is described where n-doped polysilicon thin films are used as piezoresistances, on top of membrane made of a stack of silicon dioxide/silicon nitride layers. Draw the process flow and estimate the resistivity of the polysilicon layer to achieve 4.6 kΩ for a width $W = 20$ μm. The polysilicon strip is $t = 0.35$ μm thick and $L = 800$ μm long.

9.10 In [72] the piezoresistive and thermoelectric effects of carbon nanotubes are measured. To do so, a silicon nitride film is deposited on a bare silicon wafer. A layer of chromium/gold is patterned to provide measurement electrodes. A forest film of single-wall nanotubes is then deposited, partially covering the electrodes. Draw the process flow schematically.

9.11 In [40] a resonating gravimetric sensor fabrication process is described. The main properties of this device were analysed in Problem 6.4. Draw the process flow as a combination of silicon on glass and silicon on insulator techniques.

Appendices

A

Chapter 1 Solutions

Problem 1.1

Calculate and plot the elastic restoring force of a cantilever having width $W = 10\,\mu m$ and thickness $h = 3\,\mu m$ and for lengths from $20\,\mu m$ to $2000\,\mu m$, when the deflection at the tip is 10% of the length. Take $E = 164 \times 10^9$ Pa.

Taking into account that

$$F = \frac{EWt^3}{4L^3}\delta \tag{A.1}$$

and

$$\delta = \frac{L}{10} \tag{A.2}$$

$$F = \frac{EWt^3}{40L^2}. \tag{A.3}$$

As can be seen in Figure A.1, the elastic restoring force expands over several decades with a slope of -2 in the log–log plot according to equation (A.3).

Problem 1.2

For a silicon cantilever such as the one depicted in Figure 1.1, what is the most effective way to reduce the elastic constant k by a factor of 10 by changing only one of the dimensions? Similarly, what is the most effective way to increase k by a factor of 10?

Understanding MEMS: Principles and Applications, First Edition. Luis Castañer.
© 2016 John Wiley & Sons, Ltd. Published 2016 by John Wiley & Sons, Ltd.
Companion Website: www.wiley.com/go/castaner/understandingmems

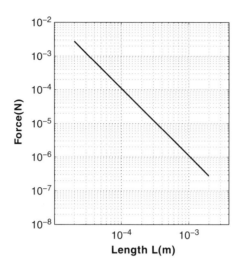

Figure A.1 Problem 1.1

To reduce (increase) the elastic constant by a factor of 10 we can reduce (increase) the thickness t by a factor of $\sqrt[3]{10}$ or increase (reduce) the length L, again by a factor of $\sqrt[3]{10}$.

Problem 1.3

We have an accelerometer based on an inertial mass from a cubic volume of silicon of 500 μm side. Find the density of silicon and calculate the force that creates such mass when accelerated at 60 times gravity. If the inertial mass is supported by a flexure having an elastic constant of 100 N/m, find the mass displacement when the two forces reach equilibrium. If the edge of the cubic volume is at 5 μm distance of a fixed electrode, find the capacitance value before the acceleration is applied to the mass and after. Assume that there is air in between the plates.

The density of silicon is 2339 kg/m^3. The force is the product of the mass and the acceleration, so

$$F = (500 \times 10^{-6})^3 \times 2339 \times 60 \times 9.8 = 1.71 \times 10^{-4}\,\text{N}. \tag{A.4}$$

The equilibrium is

$$F = k\delta = 1.71 \times 10^{-4}, \tag{A.5}$$

and hence

$$\delta = \frac{F}{k} = \frac{1.71 \times 10^{-4}}{100} = 1.71 \times 10^{-6}. \tag{A.6}$$

When there is no displacement of the mass,

$$C_0 = \frac{\epsilon_0 A}{g_0} = \frac{8.85 \times 10^{-12}(500 \times 10^{-6})^2}{5 \times 10^{-6}} = 4.42 \times 10^{-13}\ F, \tag{A.7}$$

and when the gap is reduced to $g_0 - \delta$,

$$C = \frac{\epsilon_0 A}{g_0 - \delta} = \frac{8.85 \times 10^{-12}(500 \times 10^{-6})^2}{5 \times 10^{-6} - 1.71 \times 10^{-6}} = 6.72 \times 10^{-13}\ F. \tag{A.8}$$

Problem 1.4

We have two plates of silver of area 250 μm × 250 μm and 50 μm thick. The upper plate is fixed and the bottom plate can move vertically. Calculate the minimum voltage that should be applied between the plates in order to start lifting the bottom plate.

The only force opposing the lifting of the bottom plate is its own weight, so

$$F = (250 \times 10^{-6})^2 \times 50 \times 10^{-6} \times 2339 \times 9.8 = 7.16 \times 10^{-8}\ N. \tag{A.9}$$

From the equation for electrostatic force,

$$F = \frac{\epsilon A V^2}{2g^2}, \tag{A.10}$$

we can isolate the voltage,

$$V = g\sqrt{\frac{2F}{\epsilon A}} = 2.54\ V, \tag{A.11}$$

with $A = (250 \times 10^{-6})^2\ m^2$.

Problem 1.5

We have two parallel electrodes at a distance of 4 μm in air, with a voltage $V_{CC} = 10V$ applied between them. The permittivity ϵ_1 of air is equal to the permittivity of the vacuum ϵ_0. One of the electrodes is covered by a dielectric 2 μm thick having a permittivity of $\epsilon_2 = \epsilon_0 \epsilon_r$ with $\epsilon_r = 3.9$. Calculate the electric field in the air and inside the dielectric.

We first apply Gauss's law to the boundary surface between the air and the dielectric. According to the signs shown in Figure A.2, on the right-hand side the electric field and the differential

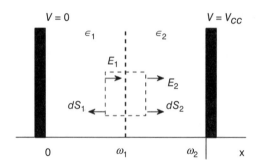

Figure A.2   Problem 1.5

of the surface have the same sign (the sign convention is that the differential of the surface of a volume is positive when directed outwards from the volume) whereas on the left-hand side they have opposite signs. We have

$$\oint DdS = -\int \epsilon_1 E_1 dS_1 + \int \epsilon_2 E_2 dS_2 = 0, \tag{A.12}$$

as there is no charge at the interface. Then

$$-\epsilon_1 E_1 + \epsilon_2 E_2 = 0. \tag{A.13}$$

Next we integrate the electric field to find the potential. For $0 < x < w_1$,

$$V(x) = -\int_0^x Edx = -E_1 x + C_1, \tag{A.14}$$

and for $w_1 < x < w_2$,

$$V(x) = -\int_{w_1}^x Edx = -E_2(x - w_1) + C_2, \tag{A.15}$$

where C_1 and C_2 are integration constants. We have four unknowns, E_1, E_2, C_1 and C_2, and we need four boundary conditions. At $x = 0$ we will assume that we have the potential reference (or ground) $V(x = 0) = 0$. At $x = w_2$ we will assume that we have the applied potential $V(x = w_2) = V_{CC}$. At the boundary between the air and the dielectric we will have two more conditions, one the result of Gauss's law in equation (A.13) and the other the continuity of the potential. So

$$V(0) = C_1 = 0, \tag{A.16}$$

$$V(w_2) = -E_2(w_2 - w_1) + C_2 = V_{CC}, \tag{A.17}$$

$$V(w_1) = -E_1 w_1 + C_1 = C_2, \tag{A.18}$$

From these equations we get

$$E_1 = \frac{-V_{CC}}{\frac{\epsilon_1}{\epsilon_2}(w_2 - w_1) + w_1} \quad \text{and} \quad E_2 = \frac{-V_{CC}}{(w_2 - w_1) + \frac{\epsilon_2}{\epsilon_1}w_1}. \tag{A.19}$$

With the data for the problem we have that $E_1 = -3.97 \times 10^6$ V/m and $E_2 = -1.01 \times 10^6$ V/m.

Problem 1.6

A thin cylindrical capillary of 5 mm diameter is immersed in water. The surface tension is $\gamma = 72.8 \times 10^{-3}$ N/m, the liquid density is $\rho_m = 10^3$ kg/m^3 and the acceleration due to gravity is $g = 9.8$ m/s^2. Calculate the height of the water inside.

The vector force F is the product of the surface tension (which is the force per unit length) multiplied by the perimeter of the liquid,

$$F = \gamma 2\pi r. \tag{A.20}$$

If the contact angle is θ, the upward force is

$$F_{\text{up}} = F \cos \theta = \gamma 2\pi r \cos \theta. \tag{A.21}$$

The downward force is the weight

$$F_{\text{down}} = \pi r^2 h g \rho_m. \tag{A.22}$$

Then

$$h = \frac{2\gamma \cos \theta}{\rho_m g r} = 4.2 \text{ mm}. \tag{A.23}$$

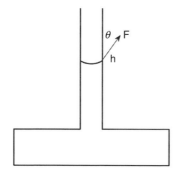

Figure A.3 Problem 1.6

Problem 1.7

Compare the surface energy of a drop of liquid, assumed spherical in shape with radius r, with its volume.

The surface energy of a drop is the drop surface multiplied by the surface tension,

$$U = 4\pi r^2 \gamma, \tag{A.24}$$

and the volume is

$$vol = \frac{4}{3}\pi r^3. \tag{A.25}$$

Then

$$\frac{U}{vol} = \frac{3\gamma}{r}. \tag{A.26}$$

As can be seen, large drops are energetically better, and this explains why smaller drops tend to form single and larger drops.

Problem 1.8

We have a spherical drop of 2 mm radius. If the surface tension is $\gamma = 72.8 \times 10^{-3}$ N/m, calculate the surface energy change if the drop radius is stretched by Δr and find the change in the internal pressure in equilibrium.

The surface energy of the drop before stretching is

$$U = 4\pi r^2 \gamma. \tag{A.27}$$

The change in energy after stretching is

$$dU = \frac{dU}{dr}dr = 8\pi r\gamma dr. \tag{A.28}$$

This change has to be compensated by an increase in the pressure inside the drop ΔP. Such increase leads to an energy change of

$$\Delta P 4\pi r^2 dr, \tag{A.29}$$

which has to units of energy. In equilibrium

$$\Delta P 4\pi r^2 dr = 8\pi r\gamma dr, \tag{A.30}$$

and the pressure change is given by

$$\Delta P = \frac{2\gamma}{r} = 72.8\,\text{Pa}. \tag{A.31}$$

Problem 1.9

A piezoelectric actuator has to produce a displacement in the bottom plate of a reservoir to eject droplets of ink to produce 600 dots per inch. Assume that the ink dot thickness is 1 μm and that there is just one drop per dot. Calculate the diameter of the dot, the volume of the drop and the radius of the drop (assumed to be equal to the radius of the ejecting nozzle). Calculate the vertical expansion required for a piezoelectric actuator acting on a cylindrical ink reservoir of 2 mm diameter. The thickness of the piezoelectric material is 10 μm.

For a circular dot of radius a, the value of a can be calculated from the number of dots per inch,

$$2a \times 600 = 1 \text{ inch} = 2.54 \times 10^{-2}\,\text{m}, \tag{A.32}$$

that is,

$$a = \frac{2.54 \times 10^{-2}}{2 \times 600} = 21.1 \times 10^{-6}\,\text{m}. \tag{A.33}$$

Neglecting any loss of volume, the volume of the dot must be equal to the volume of the drop. If the radius of the drop is r and the thickness of the dot is h (Figure A.4), we can write

$$\frac{4}{3}\pi r^3 = \pi a^2 h, \tag{A.34}$$

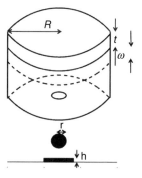

Figure A.4   Problem 1.9

so that

$$r = \left(\frac{3\pi a^2 h}{4} \right)^{1/3} = 10.1 \ \mu m. \tag{A.35}$$

The volume of the drop has to be equal to the volume displaced by the actuator, which equals the area of the reservoir multiplied by the displacement, w:

$$\pi R^2 w = \frac{4}{3} \pi r^3, \tag{A.36}$$

so

$$w = \frac{4r^3}{3R^2} = 1.37 \times 10^{-9} \ m. \tag{A.37}$$

If the actuator has a thickness of $t = 10 \ \mu m$ and has to produce a deformation of w, the strain S is

$$S = \frac{w}{t} = 1.37 \times 10^{-4}. \tag{A.38}$$

The strain is related to the electric field by means of the piezoelectric coefficient d. Assuming that $d = 480 \times 10^{-12}$, the electric field is

$$E = \frac{S}{d} = 2.86 \times 10^5 \ V/m. \tag{A.39}$$

The voltage that should be applied across the 10 μm thick piezo material is then

$$V = E \times t = 2.86 \ V. \tag{A.40}$$

Problem 1.10

We have a flexure made of gold and an electrical current of 10 mA circulates through it. If we immerse the flexure in a magnetic field normal to the plane of the flexure, calculate the force and indicate the direction of the movement. Take L = 2000 μm, B = 0.25 × 10⁻⁴ T and the elastic constant of the flexure k_x = 0.01 N/m.

In Figure A.5 there are three fixed supports. The top and bottom ones support two flexures of smaller thickness that hold an inertial mass thicker than the flexures. On the right there is a third fixed electrode that will sense the deflection in the x direction as shown. It the magnetic field has the direction show by the arrow, and the electric current the direction also shown, then the movement will be in the direction of the x-axis. The Lorentz force is

$$F = I\vec{L} \times \vec{B}. \tag{A.41}$$

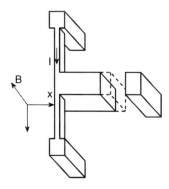

Figure A.5   Problem 1.10

As the wire and the magnetic field are orthogonal, the force is directly given by

$$F = ILB = 1 \times 10^{-9}\,\text{N},\tag{A.42}$$

and the displacement is

$$\delta = \frac{F}{k_x} = 0.1\,\mu\text{m}.\tag{A.43}$$

Problem 1.11

We have a 300 nm diameter polystyrene nanoparticle immersed in air, and an electric field created by a sphere carrying a total charge of Q. The distance between the particle and the sphere is 10 times the polystyrene sphere diameter. Calculate the force and the direction. Repeat the calculation if the medium is changed to ethylene glycol. Ethylene glycol has a relative permittivity of 37 and polystyrene 2.5.

The electric field created by a sphere with a charge Q is

$$E = \frac{Q}{r^2},\tag{A.44}$$

and has radial symmetry. The modulus squared of the field is

$$|E|^2 = \frac{Q^2}{r^4},\tag{A.45}$$

and the gradient is

$$\nabla |E|^2 = -4\frac{Q^2}{r^5}.\tag{A.46}$$

The dielectrophoretic force is given by

$$\vec{F} = 2\pi\epsilon_1 R^3 K\nabla|E|^2, \tag{A.47}$$

where K is the Clausius–Mossotti factor

$$K = \frac{\epsilon_2 - \epsilon_1}{\epsilon_2 + 2\epsilon_1}, \tag{A.48}$$

in which ϵ_1 is the fluid permittivity and ϵ_2 is the particle permittivity. Substituting yields

$$F = -8\pi\epsilon_1 R^3 \frac{\epsilon_2 - \epsilon_1}{\epsilon_2 + 2\epsilon_1} \frac{Q^2}{r^5}. \tag{A.49}$$

When the medium is air the Clausius–Mossotti factor is

$$K = \frac{2.5 - 1}{2.5 + 2} = 0.333. \tag{A.50}$$

At a distance 10 times the radius of the particle, $r = 10R$, we assume that the gradient of the electric field squared is 3.3×10^{23} V^2/m^3, the force is $F = -6.59 \times 10^{-9}$ N and the sign is attractive towards the charged sphere. If the medium is ethylene glycol the value of the Clausius–Mossotti factor is

$$K = \frac{2.5 - 37}{2.5 + 74} = -0.45 \tag{A.51}$$

and then the force is $F = 8.9 \times 10^{-9}$ N and is repulsive.

B

Chapter 2 Solutions

Problem 2.1

Find suitable axes for a silicon wafer with orientation (100), and demonstrate that the three unit vectors are orthogonal.

A silicon wafer (100) indicates that the plane of the wafer is one of the planes equivalent to (100), and we can still use our definition of the main crystallographic axes. The flat orientation of the wafer that is aligned to the direction [110], and then a suitable set of axes is X' ([110]), Y' ([1$\bar{1}$0]) and Z' ([001]).

It is easy to show that the three rotated axes are orthogonal:

$$\cos\theta_{\vec{i_1'}\vec{i_2'}} = \frac{-1+1+0}{\sqrt{2}\sqrt{2}} = 0, \tag{B.1}$$

$$\cos\theta_{\vec{i_1'}\vec{i_3'}} = \frac{0+0+0}{\sqrt{2}\sqrt{1}} = 0, \tag{B.2}$$

$$\cos\theta_{\vec{i_2'}\vec{i_3'}} = \frac{0+0+0}{\sqrt{2}\sqrt{1}} = 0. \tag{B.3}$$

Problem 2.2

Using the set of unit vectors found in Problem 2.1, calculate the rotation matrix from the crystallographic axes to the new ones.

Understanding MEMS: Principles and Applications, First Edition. Luis Castañer.
© 2016 John Wiley & Sons, Ltd. Published 2016 by John Wiley & Sons, Ltd.
Companion Website: www.wiley.com/go/castaner/understandingmems

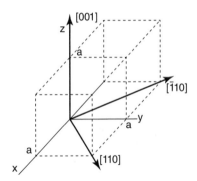

Figure B.1   Axes of a silicon wafer (100)

The rotation matrix between the set of axis (XYZ), $\vec{i}_1 = [100]$, $\vec{i}_2 = [010]$ and $\vec{i}_3 = [001]$, and $(X'Y'Z')$, $\vec{i}_1' = [110]$, $\vec{i}_2' = [\bar{1}10]$, and $\vec{i}_3' = [001]$, is given by

$$(R) = \begin{pmatrix} \cos\theta_{i_1'i_1} & \cos\theta_{i_1'i_2} & \cos\theta_{i_1'i_3} \\ \cos\theta_{i_2'i_1} & \cos\theta_{i_2'i_2} & \cos\theta_{i_2'i_3} \\ \cos\theta_{i_3'i_1} & \cos\theta_{i_3'i_2} & \cos\theta_{i_3'i_3} \end{pmatrix} = \begin{pmatrix} l_1 & m_1 & n_1 \\ l_2 & m_2 & n_2 \\ l_3 & m_3 & n_3 \end{pmatrix} = \begin{pmatrix} \frac{1}{\sqrt{2}} & \frac{1}{\sqrt{2}} & 0 \\ \frac{-1}{\sqrt{2}} & \frac{1}{\sqrt{2}} & 0 \\ 0 & 0 & 1 \end{pmatrix}. \tag{B.4}$$

Problem 2.3

Using the set of unit vectors found in Problem 2.1, calculate the values of Young's modulus and the Poisson ratio in the rotated axes.

We consider the X'-axis as the longitudinal axis and the Y'-axis as the transversal axis – these are the two axes located in the plane of the wafer. We have

$$E_{110} = \frac{T_1'}{S_1'} = \frac{1}{s_{11}'}, \tag{B.5}$$

and

$$s_{11}' = s_{11} + \left(s_{11} - s_{12} - \frac{1}{2}s_{44} \right) \left(l_1^4 + m_1^4 + n_1^4 - 1 \right), \tag{B.6}$$

with $s_{11} = 0.0077$ GPa^{-1}, $s_{12} = 0.0021$ GPa^{-1} and $s_{44} = 0.0125$ GPa^{-1}, so

$$E_{110} = \frac{1}{s_{11}'} = \frac{1}{6.81 \times 10^{-3}} = 168.7 \, \text{GPa}. \tag{B.7}$$

The Poisson ratio is defined similarly to equation (2.60) in the rotated axes, as the longitudinal direction is assigned to the X'-axis and the transversal direction to the Y'-axis,

$$v_{110} = -\frac{S_2'}{S_1'} = -\frac{s_{12}'}{s_{11}'}, \tag{B.8}$$

with

$$s_{12}' = s_{12} + \left(s_{11} - s_{12} - \frac{1}{2}s_{44}\right)(l_1^2 l_2^2 + m_1^2 m_2^2 + n_1^2 n_2^2) = -3.25 \times 10^{-4}, \tag{B.9}$$

so

$$v_{110} = -\frac{s_{12}'}{s_{11}'} = 0.054. \tag{B.10}$$

The same result is obtained if we consider the Y'-axis as the longitudinal axis and the X'-axis as the transversal axis, as then the only component of the stress different from zero is T_2' along the Y'-axis, so that

$$S_1' = s_{12}' T_2' \tag{B.11}$$

and

$$S_2' = s_{11}' T_2'. \tag{B.12}$$

Then

$$E_{\bar{1}10} = \frac{1}{s_{11}'} = \frac{1}{6.81 \times 10^{-3}} = 168.7 \, \text{GPa}, \tag{B.13}$$

and similarly

$$v_{\bar{1}10} = -\frac{s_{12}'}{s_{11}'} = 0.054. \tag{B.14}$$

Problem 2.4

Using the set of unit vectors found in Problem 2.1, calculate the values of the elements C_{11}' and C_{12}' of the stiffness matrix in the rotated axes.

We first calculate the value of C_c:

$$C_c = C_{11} - C_{12} - 2C_{44} = 166 - 64 - 160 = -50 \, \text{GPa}. \tag{B.15}$$

Then

$$C'_{11} = C_{11} + C_c \left(l_1^4 + m_1^4 + n_1^4 - 1 \right) = 166 - 58 \left(\frac{1}{4} + \frac{1}{4} - 1 \right) = 195 \, \text{GPa},$$

$$C'_{12} = C_{12} + C_c \left(l_1^2 l_2^2 + m_1^2 m_2^2 + n_1^2 n_2^2 \right) = 64 - 58 \left(\frac{1}{4} + \frac{1}{4} \right) = 35 \, \text{GPa}.$$

Problem 2.5

Write Matlab code and find the values of all elements of the stiffness matrix in the rotated system for Problem 2.1.

The Matlab code is as follows:

```
%Stiffness matrix rotation
%Enter the subindices of the stiffness element
i=1
j=1
k=1
l=1
%Enter the rotation matrix, from problem 2.2
beta=45*pi/180
%Rotation matrix around axis z with angle beta
R=[[0.707 0.707 0]
   [-0.707 0.707 0]
   [0 0 1]
   ]
%Stiffness matrix in crystallographic axes in short notation
Cshort=[[166 64 64 0 0 0]

       [64 166 64 0 0 0]
       [64 64 166 0 0 0]
       [0 0 0 80 0 0]
       [0 0 0 0 80 0]
       [0 0 0 0 0 80]
       ]
% Expanded indices notation of the unrotated stiffness matrix elements
C(1,1,1,1)=Cshort(1,1);
C(1,1,1,2)=Cshort(1,6);
C(1,1,1,3)=Cshort(1,5);
C(1,1,2,1)=Cshort(1,6);
C(1,1,2,2)=Cshort(1,2);
C(1,1,2,3)=Cshort(1,4);
C(1,1,3,1)=Cshort(1,5);
C(1,1,3,2)=Cshort(1,4);
```

```
C(1,1,3,3)=Cshort(1,3);

C(1,2,1,1)=Cshort(6,1);
C(1,2,1,2)=Cshort(6,6);
C(1,2,1,3)=Cshort(6,5);
C(1,2,2,1)=Cshort(6,6);
C(1,2,2,2)=Cshort(6,2);
C(1,2,2,3)=Cshort(6,4);
C(1,2,3,1)=Cshort(6,5);
C(1,2,3,2)=Cshort(6,4);
C(1,2,3,3)=Cshort(6,3);

C(1,3,1,1)=Cshort(5,1);
C(1,3,1,2)=Cshort(5,6);
C(1,3,1,3)=Cshort(5,5);
C(1,3,2,1)=Cshort(5,6);
C(1,3,2,2)=Cshort(5,2);
C(1,3,2,3)=Cshort(5,4);
C(1,3,3,1)=Cshort(5,5);
C(1,3,3,2)=Cshort(5,4);
C(1,3,3,3)=Cshort(5,3);

C(2,1,1,1)=Cshort(6,1);
C(2,1,1,2)=Cshort(6,6);
C(2,1,1,3)=Cshort(6,5);
C(2,1,2,1)=Cshort(6,6);
C(2,1,2,2)=Cshort(6,2);
C(2,1,2,3)=Cshort(6,4);
C(2,1,3,1)=Cshort(6,5);
C(2,1,3,2)=Cshort(6,4);
C(2,1,3,3)=Cshort(6,3);

C(2,2,1,1)=Cshort(2,1);
C(2,2,1,2)=Cshort(2,6);
C(2,2,1,3)=Cshort(2,5);
C(2,2,2,1)=Cshort(2,6);
C(2,2,2,2)=Cshort(2,2);
C(2,2,2,3)=Cshort(2,4);
C(2,2,3,1)=Cshort(2,5);
C(2,2,3,2)=Cshort(2,4);
C(2,2,3,3)=Cshort(2,3);

C(2,3,1,1)=Cshort(4,1);
C(2,3,1,2)=Cshort(4,6);
C(2,3,1,3)=Cshort(4,5);
C(2,3,2,1)=Cshort(4,6);
C(2,3,2,2)=Cshort(4,2);
```

```
C(2,3,2,3)=Cshort(4,4);
C(2,3,3,1)=Cshort(4,5);
C(2,3,3,2)=Cshort(4,4);
C(2,3,3,3)=Cshort(4,3);

C(3,1,1,1)=Cshort(5,1);
C(3,1,1,2)=Cshort(5,6);
C(3,1,1,3)=Cshort(5,5);
C(3,1,2,1)=Cshort(5,6);
C(3,1,2,2)=Cshort(5,2);
C(3,1,2,3)=Cshort(5,4);
C(3,1,3,1)=Cshort(5,5);
C(3,1,3,2)=Cshort(5,4);
C(3,1,3,3)=Cshort(5,3);

C(3,2,1,1)=Cshort(4,1);
C(3,2,1,2)=Cshort(4,6);
C(3,2,1,3)=Cshort(4,5);
C(3,2,2,1)=Cshort(4,6);
C(3,2,2,2)=Cshort(4,2);
C(3,2,2,3)=Cshort(4,4);
C(3,2,3,1)=Cshort(4,5);
C(3,2,3,2)=Cshort(4,4);
C(3,2,3,3)=Cshort(4,3);

C(3,3,1,1)=Cshort(3,1);
C(3,3,1,2)=Cshort(3,6);
C(3,3,1,3)=Cshort(3,5);
C(3,3,2,1)=Cshort(3,6);
C(3,3,2,2)=Cshort(3,2);
C(3,3,2,3)=Cshort(3,4);
C(3,3,3,1)=Cshort(3,5);
C(3,3,3,2)=Cshort(3,4);
C(3,3,3,3)=Cshort(3,3);
% Calculation of the element of the rotated matrix
cprime=0;

for a=1:3
    for b=1:3
        for c=1:3
            for d=1:3
                cprime=cprime+R(i,a)*R(j,b)*R(k,c)*R(l,d)*C(a,b,c,d);
            end
        end
    end
end
result=cprime
```

Running this code for the 6×6 elements of the rotated matrix yields

$$(C') = \begin{pmatrix} 195 & 35 & 64 & 0 & 0 & 0 \\ 35 & 195 & 64 & 0 & 0 & 0 \\ 64 & 64 & 166 & 0 & 0 & 0 \\ 0 & 0 & 0 & 80 & 0 & 0 \\ 0 & 0 & 0 & 0 & 80 & 0 \\ 0 & 0 & 0 & 0 & 0 & 51 \end{pmatrix}. \tag{B.16}$$

Problem 2.6

Using the set of unit vectors found in Problem 2.1, calculate the values of the elements of the compliance matrix in the rotated axes.

The Matlab code

```
cprime=[195 35 64 0 0 0; 35 195 64 0 0 0; 64 64 166 0 0 0; 0 0 0 80
0 0; 0 0 0 0 80 0; 0 0 0 0 0 51]
sprime=inv (cprime)
```

calculates the compliance matrix by inverting the stiffness matrix, and the result is

$$(s') = \begin{pmatrix} 0.0059 & -0.0004 & -0.0021 & 0 & 0 & 0 \\ -0.0004 & 0.0059 & -0.0021 & 0 & 0 & 0 \\ -0.0021 & -0.0021 & 0.0077 & 0 & 0 & 0 \\ 0 & 0 & 0 & 0.0125 & 0 & 0 \\ 0 & 0 & 0 & 0 & 0.0125 & 0 \\ 0 & 0 & 0 & 0 & 0 & 0.0196 \end{pmatrix}. \tag{B.17}$$

Problem 2.7

Show that the element C'_{12} of the stiffness matrix in the rotated axes is given by [12]

$$C'_{12} = C_{12} + (C_{11} - C_{12} - 2C_{44})(l_1^2 l_2^2 + m_1^2 m_2^2 + n_1^2 n_2^2). \tag{B.18}$$

The starting equation is

$$c'_{iljk} = \sum_{a=1}^{a=3} \sum_{b=1}^{b=3} \sum_{c=1}^{c=3} \sum_{d=1}^{d=3} a_{ia} a_{lb} a_{jc} a_{kd} C_{abcd}, \tag{B.19}$$

and expanding the subindices of $C'_{12} \to C'_{1122}$ yields

$$c'_{1122} = \sum_{a=1}^{a=3} \sum_{b=1}^{b=3} \sum_{c=1}^{c=3} \sum_{d=1}^{d=3} a_{1a} a_{1b} a_{2c} a_{2d} C_{abcd}. \tag{B.20}$$

Taking into account that $C_{14} = C_{1123} = 0, C_{15} = C_{1113} = 0, C_{16} = C_{1112} = 0, C_{24} = C_{2223} = 0, C_{25} = C_{2213} = 0, C_{26} = C_{2212} = 0, C_{34} = C_{3323} = 0, C_{35} = C_{3313} = 0, C_{36} = C_{3312} = 0, C_{41} = C_{2311} = 0, C_{42} = C_{2322} = 0, C_{43} = C_{2333} = 0, C_{45} = C_{2313} = 0, C_{46} = C_{2312} = 0, C_{51} = C_{1211} = 0, C_{52} = C_{1322} = 0, C_{53} = C_{1333} = 0, C_{54} = C_{1323} = 0, C_{56} = C_{1312} = 0, C_{61} = C_{1211} = 0, C_{62} = C_{1222} = 0, C_{63} = C_{1233} = 0, C_{64} = C_{1223} = 0, C_{65} = C_{1213} = 0$, we find that

$$C'_{12} = C_{11}\left(l_1^2 l_2^2 + m_1^2 m_2^2 + n_1^2 n_2^2\right) + C_{12}\left(l_1^2\left(m_2^2 + n_2^2\right) + m_1^2\left(l_2^2 + n_2^2\right) + n_1^2\left(l_2^2 + m_2^2\right)\right)$$
$$+ 4C_{44}(l_1 m_1 l_2 m_2 + l_1 n_1 l_2 n_2 + m_1 n_1 m_2 n_2).$$

We know that the director cosines obey the equation

$$l_1 l_2 + m_1 m_2 + n_1 n_2 = 0. \tag{B.21}$$

Then

$$(l_1 l_2 + m_1 m_2 + n_1 n_2)^2 = 0 \tag{B.22}$$

and

$$(l_1 l_2 + m_1 m_2 + n_1 n_2)^2 = l_1^2 l_2^2 + m_1^2 m_2^2 + n_1^2 n_2^2 + \tag{B.23}$$
$$+ 2(l_1 l_2 m_1 m_2 + l_1 l_2 n_1 n_2 + m_1 m_2 n_1 n_2) = 0,$$

together with equation (B.21),

$$C'_{12} = (C_{11} - 2C_{44})\left(l_1^2 l_2^2 + m_1^2 m_2^2 + n_1^2 n_2^2\right) + C_{12}\left(l_1^2\left(m_2^2 + n_2^2\right) + m_1^2\left(l_2^2 + n_2^2\right) + n_1^2\left(l_2^2 + m_2^2\right)\right), \tag{B.24}$$

Taking into account that the direction cosines also satisfy the equation

$$l_2^2 + m_2^2 + n_2^2 = 1, \tag{B.25}$$

and then

$$m_2^2 + n_2^2 = 1 - l_2^2,$$
$$l_2^2 + n_2^2 = 1 - m_2^2,$$
$$l_2^2 + m_2^2 = 1 - n_2^2, \tag{B.26}$$

substituting in equation (B.24) leads to

$$C'_{12} = C_{12} + (C_{11} - C_{12} - 2C_{44})\left(l_1^2 l_2^2 + m_1^2 m_2^2 + n_1^2 n_2^2\right). \tag{B.27}$$

Problem 2.8

Calculate the rotation matrix when the crystallographic axes are rotated around the z-axis by an angle β.

The original unit vector basis is $\vec{i}_1 = (100)$, $\vec{i}_2 = (010)$ and $\vec{i}_3 = (001)$. In the rotation the axis $\vec{i}_3$ remains the same, $\vec{i}'_3 = \vec{i}_3$, so the angles between the axes $\vec{i}'_1$ and $\vec{i}_3$, and between $\vec{i}'_2$ and $\vec{i}_3$, are $\pi/2$. According to Figure B.2, the director cosines are

$$\cos\left(i_1 i'_1\right) = \cos\left(i_2 i'_2\right) = \cos\beta, \tag{B.28}$$

$$\cos\left(i_1 i'_2\right) = \cos\left(\frac{\pi}{2} + \beta\right) = -\sin\beta, \tag{B.29}$$

$$\cos\left(i_2 i'_1\right) = \cos\left(\frac{\pi}{2} - \beta\right) = \sin\beta, \tag{B.30}$$

and thus the rotation matrix is

$$(R) = \begin{pmatrix} \cos\beta & \sin\beta & 0 \\ -\sin\beta & \cos\beta & 0 \\ 0 & 0 & 1 \end{pmatrix}. \tag{B.31}$$

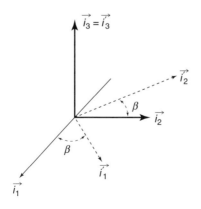

Figure B.2   Problem 2.8

Problem 2.9

Write Matlab code or similar to calculate the values of the elements C'_{11} and C'_{12} of the rotated stiffness matrix when the rotation is around the z-axis and with angles of 15°, 30°, 45°, 60° and 90°.

Running the following code yields the results in Table B.1.

```
%Stiffness matrix rotation
%Enter the subindices of
i=1
j=1
k=1
l=1
%Enter the angle
beta=15*pi/180
%Rotation matrix around axis z with angle beta
R=[[cos(beta) sin(beta) 0]
    [-sin(beta) cos(beta) 0]
    [0 0 1]
    ]
%Stiffness matrix in crystallographic axes in short notation
Cshort=[[166 64 64 0 0 0]

    [64 166 64 0 0 0]
    [64 64 166 0 0 0]
    [0 0 0 80 0 0]
    [0 0 0 0 80 0]
    [0 0 0 0 0 80]
    ]
% Expanded indices notation of the unrotated stiffness matrix elements
C(1,1,1,1)=Cshort(1,1);
```

Table B.1 Results for Problem 2.9

β	C'_{1111} (GPa)	C'_{1122} (GPa)
0	166	64
15	173.25	56.75
30	187.75	42.25
45	195	35
60	187.75	42.25
90	166	64

```
C(1,1,1,2)=Cshort(1,6);
C(1,1,1,3)=Cshort(1,5);
C(1,1,2,1)=Cshort(1,6);
C(1,1,2,2)=Cshort(1,2);
C(1,1,2,3)=Cshort(1,4);
C(1,1,3,1)=Cshort(1,5);
C(1,1,3,2)=Cshort(1,4);
C(1,1,3,3)=Cshort(1,3);

C(1,2,1,1)=Cshort(6,1);
C(1,2,1,2)=Cshort(6,6);
C(1,2,1,3)=Cshort(6,5);
C(1,2,2,1)=Cshort(6,6);
C(1,2,2,2)=Cshort(6,2);
C(1,2,2,3)=Cshort(6,4);
C(1,2,3,1)=Cshort(6,5);
C(1,2,3,2)=Cshort(6,4);
C(1,2,3,3)=Cshort(6,3);

C(1,3,1,1)=Cshort(5,1);
C(1,3,1,2)=Cshort(5,6);
C(1,3,1,3)=Cshort(5,5);
C(1,3,2,1)=Cshort(5,6);
C(1,3,2,2)=Cshort(5,2);
C(1,3,2,3)=Cshort(5,4);
C(1,3,3,1)=Cshort(5,5);
C(1,3,3,2)=Cshort(5,4);
C(1,3,3,3)=Cshort(5,3);

C(2,1,1,1)=Cshort(6,1);
C(2,1,1,2)=Cshort(6,6);
C(2,1,1,3)=Cshort(6,5);
C(2,1,2,1)=Cshort(6,6);
C(2,1,2,2)=Cshort(6,2);
C(2,1,2,3)=Cshort(6,4);
C(2,1,3,1)=Cshort(6,5);
C(2,1,3,2)=Cshort(6,4);
C(2,1,3,3)=Cshort(6,3);

C(2,2,1,1)=Cshort(2,1);
C(2,2,1,2)=Cshort(2,6);
C(2,2,1,3)=Cshort(2,5);
C(2,2,2,1)=Cshort(2,6);
C(2,2,2,2)=Cshort(2,2);
C(2,2,2,3)=Cshort(2,4);
C(2,2,3,1)=Cshort(2,5);
```

```
C(2,2,3,2)=Cshort(2,4);
C(2,2,3,3)=Cshort(2,3);

C(2,3,1,1)=Cshort(4,1);
C(2,3,1,2)=Cshort(4,6);
C(2,3,1,3)=Cshort(4,5);
C(2,3,2,1)=Cshort(4,6);
C(2,3,2,2)=Cshort(4,2);
C(2,3,2,3)=Cshort(4,4);
C(2,3,3,1)=Cshort(4,5);
C(2,3,3,2)=Cshort(4,4);
C(2,3,3,3)=Cshort(4,3);

C(3,1,1,1)=Cshort(5,1);
C(3,1,1,2)=Cshort(5,6);
C(3,1,1,3)=Cshort(5,5);
C(3,1,2,1)=Cshort(5,6);
C(3,1,2,2)=Cshort(5,2);
C(3,1,2,3)=Cshort(5,4);
C(3,1,3,1)=Cshort(5,5);
C(3,1,3,2)=Cshort(5,4);
C(3,1,3,3)=Cshort(5,3);

C(3,2,1,1)=Cshort(4,1);
C(3,2,1,2)=Cshort(4,6);
C(3,2,1,3)=Cshort(4,5);
C(3,2,2,1)=Cshort(4,6);
C(3,2,2,2)=Cshort(4,2);
C(3,2,2,3)=Cshort(4,4);
C(3,2,3,1)=Cshort(4,5);
C(3,2,3,2)=Cshort(4,4);
C(3,2,3,3)=Cshort(4,3);

C(3,3,1,1)=Cshort(3,1);
C(3,3,1,2)=Cshort(3,6);
C(3,3,1,3)=Cshort(3,5);
C(3,3,2,1)=Cshort(3,6);
C(3,3,2,2)=Cshort(3,2);
C(3,3,2,3)=Cshort(3,4);
C(3,3,3,1)=Cshort(3,5);
C(3,3,3,2)=Cshort(3,4);
C(3,3,3,3)=Cshort(3,3);
% Calculation of the element of the rotated matrix
cprime=0;

for a=1:3
```

```
    for b=1:3
        for c=1:3
            for d=1:3
                cprime=cprime+R(i,a)*R(j,b)*R(k,c)*R(l,d)*C(a,b,c,d);
            end
        end
    end
end

result=cprime
```

Problem 2.10

Calculate the values of Young's modulus and the Poisson ratio in the plane when the rotation is around the z-axis and with angles of $15°$, $30°$ *and* $45°$.

We run the Matlab code to find the rotated stiffness matrix and then we invert it to find the rotated compliance matrix for every angle.

For $\beta = 15°$,

$$
(C'(15°)) = \begin{pmatrix}
173.2 & 56.7 & 64 & 0 & 12.5 & 12.5 \\
56.7 & 173.2 & 64 & 0 & 0 & -12.5 \\
64 & 64 & 166 & 0 & 0 & 0 \\
0 & 0 & 0 & 80 & 0 & 0 \\
0 & 0 & 0 & 0 & 80 & 0 \\
12.5 & -12.5 & 0 & 0 & 0 & 72.7
\end{pmatrix}, \tag{B.32}
$$

$$
(s'(15°)) = \begin{pmatrix}
0.0072 & -0.0017 & -0.0021 & 0 & -0.0011 & -0.0015 \\
-0.0017 & 0.0072 & -0.0021 & 0 & -0.003 & 0.0015 \\
-0.0021 & -0.0021 & 0.0077 & 0 & 0.003 & 0 \\
0 & 0 & 0 & 0.0125 & 0 & 0 \\
0 & 0 & 0 & 0 & 0.0125 & 0 \\
-0.0015 & 0.0015 & 0 & 0 & 0.0002 & 0.0143
\end{pmatrix}. \tag{B.33}
$$

Young's modulus is given by

$$
E = \frac{1}{s'_{11}} = 138.8\,\text{GPa}, \tag{B.34}
$$

and the Poisson ratio by

$$
v = -\frac{s'_{12}}{s'_{11}} = 0.236. \tag{B.35}
$$

For $\beta = 30°$,

$$
\left(C'(30°) \right) = \begin{pmatrix}
187.7 & 42.25 & 64 & 0 & 12.5 & 12.5 \\
42.25 & 187.7 & 64 & 0 & 0 & -12.5 \\
64 & 64 & 166 & 0 & 0 & 0 \\
0 & 0 & 0 & 80 & 0 & 0 \\
0 & 0 & 0 & 0 & 80 & 0 \\
12.5 & -12.5 & 0 & 0 & 0 & 58.2
\end{pmatrix},
\tag{B.36}
$$

$$
\left(s'(30°) \right) = \begin{pmatrix}
0.0063 & -0.0008 & -0.0021 & 0 & -0.001 & -0.0015 \\
-0.0008 & 0.0063 & -0.0021 & 0 & 0.001 & 0.0015 \\
-0.0021 & -0.0021 & 0.0077 & 0 & 0.003 & 0 \\
0 & 0 & 0 & 0.0125 & 0 & 0 \\
0 & 0 & 0 & 0 & 0.0125 & 0 \\
-0.0015 & 0.0015 & 0 & 0 & 0.0002 & 0.0178
\end{pmatrix}.
\tag{B.37}
$$

Young's modulus is given by

$$
E = \frac{1}{s'_{11}} = 158.7 \, \text{GPa},
\tag{B.38}
$$

and the Poisson ratio by

$$
v = -\frac{s'_{12}}{s'_{11}} = 0.126.
\tag{B.39}
$$

For $\beta = 45°$, the results are the same as in Problem 2.5. Young's modulus is given by

$$
E = \frac{1}{s'_{11}} = 169.4 \, \text{GPa},
\tag{B.40}
$$

and the Poisson ratio by

$$
v = -\frac{s'_{12}}{s'_{11}} = 0.067.
\tag{B.41}
$$

The results are summarized in Table B.2.

Table B.2 Results Problem (2.10)

β	$E(GPa)$	v
0	129.8	0.272
15	138.8	0.236
30	158.7	0.126
45	169.4	0.067

C

Chapter 3 Solutions

Problem 3.1

Calculate the position of the neutral surface in a beam having a trapezoidal cross-section as shown in Figure C.1.

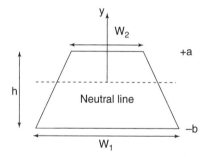

Figure C.1 Trapezoidal cross-section geometry

We have to solve the equation

$$\int_{-b}^{+a} y\,dA = 0, \tag{C.1}$$

with

$$dA = w(y)dy. \tag{C.2}$$

With the change of variable

$$u = a - y, \quad du = -dy, \tag{C.3}$$

Understanding MEMS: Principles and Applications, First Edition. Luis Castañer.
© 2016 John Wiley & Sons, Ltd. Published 2016 by John Wiley & Sons, Ltd.
Companion Website: www.wiley.com/go/castaner/understandingmems

the integration limits change as follows:

$$y = 0 \rightarrow u = a, \quad y = a \rightarrow u = 0, \quad y = -b \rightarrow u = a + b. \tag{C.4}$$

Equation (C.1) becomes

$$\int_{-b}^{+a} y \, dA = \int_{0}^{h} (a - u)w(u)du = 0, \tag{C.5}$$

as

$$\tan\theta = \frac{2h}{W_2 - W_1} = \frac{2(h - u)}{W_2 - w(u)}$$

$$w(u) = W_1 + \frac{u}{h}(W_2 - W_1) \tag{C.6}$$

$$\int_{0}^{h} (a - u)w(u)du = aW_1 h + \frac{h^2}{2}\left(\frac{a}{h}(W_2 - W_1) - W_1\right) - \frac{h^2}{3}(W_2 - W_1) = 0$$

$$a = \frac{W_1 + 2W_2}{3(W_1 + W_2)}h. \tag{C.7}$$

Problem 3.2

Calculate the moment of inertia for the geometries depicted in Figure C.2. Assume a generic volume of rectangular cross-sections in all dimensions as shown in Figure C.3.

The geometry shown in Figure C.2(a) is exactly the same as the case described in the Example 3.5.1, where the neutral plane is the (x, z) plane and the cross-section is rectangular and lies in the (y, z) plane. The bending creates a moment about the z-axis and the differential of area is $dA = bdy$:

$$I_z = \int_{-c/2}^{c/2} y^2 dA = \int_{-c/2}^{c/2} y^2 bdy = \frac{bc^3}{12}. \tag{C.8}$$

In the geometry shown in Figure C.2(b) the bending creates a moment about the y-axis, the cross-section lies in the (y, z) plane and the neutral plane is the (x, y) plane. The differential of the area of the cross-section is $dA = cdz$ and the moment of inertia I_y is given by

$$I_y = \int_{-b/2}^{b/2} z^2 dA = \int_{-b/2}^{b/2} z^2 cdz = \frac{cb^3}{12}. \tag{C.9}$$

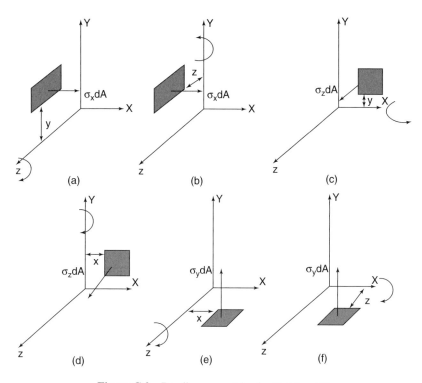

Figure C.2 Bending geometries for Problem 3.2

In the geometry shown in Figure C.2(c) the bending creates a moment about the x-axis, the cross-section lies in the (y, x) plane and the neutral plane is the (x, z) plane. The differential of the area of the cross-section is $dA = a\,dy$ and the moment of inertia I_x is given by

$$I_x = \int_{-c/2}^{c/2} y^2 dA = \int_{-c/2}^{c/2} y^2 a\,dy = \frac{ac^3}{12}. \tag{C.10}$$

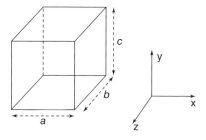

Figure C.3 Dimensions of a generic prism for Problem 3.2

In the geometry shown in Figure C.2(d) the bending creates a moment about the y-axis, the cross-section lies in the (x, y) plane and the neutral plane is the (y, z) plane. The differential of the area of the cross-section is $dA = c\,dx$ and the moment of inertia I_y is given by

$$I_y = \int_{-a/2}^{a/2} x^2 dA = \int_{-a/2}^{a/2} x^2 c\,dx = \frac{ca^3}{12}, \tag{C.11}$$

In the geometry shown in Figure C.2(e) the bending creates a moment about the z-axis the cross-section lies in the (x, z) plane and the neutral plane is the (y, z) plane. The differential of the area of the cross-section is $dA = b\,dx$ and the moment of inertia I_y is given by

$$I_z = \int_{-a/2}^{a/2} x^2 dA = \int_{-a/2}^{a/2} x^2 b\,dx = \frac{ba^3}{12}. \tag{C.12}$$

In the geometry shown in Figure C.2(f) the bending creates a moment about the x-axis, the cross-section lies in the (x, z) plane and the neutral plane is the (x, y) plane. The differential of the area of the cross-section is $dA = a\,dz$ and the moment of inertia I_y is given by

$$I_y = \int_{-b/2}^{b/2} z^2 dA = \int_{-b/2}^{b/2} z^2 a\,dz = \frac{ab^3}{12}. \tag{C.13}$$

Problem 3.3

We have a cantilever that during fabrication has accumulated a residual stress causing the upper part to be under expansion and the lower part under compression. The cross-section is rectangular, and we know that the neutral plane lies in the middle. The thickness of the cantilever is $h = 2\,\mu m$. We also know that we are in pure bending and that the value of the stress at the upper surface $\sigma_0(y = h/2) = 20\,MPa$, $W = 10\,\mu m$, $E = 160\,GPa$. Draw the stress distribution as a function of the value of the position y inside the cantilever, assuming it is linear, and write the function $\sigma(y)$. Calculate the value of bending moment, M_b, caused by the distribution of stress at the section considered. Is the bending moment clockwise or counterclockwise? Calculate the value of the radius of curvature ρ_c.

We know that the neutral axis for a rectangular cross-section lies in the middle of the beam thickness. We also know the relationship between the stress and the position y within the beam body. So

$$\sigma(y) = \frac{\sigma_0}{h/2} y = \frac{20 \times 10^6}{1 \times 10^6} y = 20y. \tag{C.14}$$

The distribution of stress in the cross-section is plotted in Figure C.4.

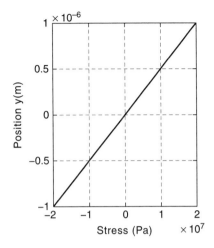

Figure C.4 Distribution of stress in the cross-section

The moment created by this distribution of stress is given by

$$M_b = - \int_A \sigma(y) y \, dA, \tag{C.15}$$

and as $dA = W dy$,

$$M_b = - \int_{-h/2}^{+h/2} W \frac{2\sigma_0}{h} y^2 \, dy = -\frac{W\sigma_0}{6h^2} = -1.33 \times 10^{-10} \, \text{N/m}. \tag{C.16}$$

The radius of curvature is given by

$$\frac{1}{\rho_c} = \frac{M_b}{EI}, \tag{C.17}$$

and with $I = Wh^3/12$,

$$\rho_c = -\frac{2\sigma_0}{Eh} = -8 \, \text{mm}. \tag{C.18}$$

The minus sign indicates that the bending is downwards.

Problem 3.4

A beam is supported at both ends by fixed supports and has a point load F at the middle x = L/2, as shown in Figure C.5. Calculate the reactions at the supports, and the deflection at any point.

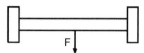

Figure C.5   Problem 3.4

The fixed supports are represented in the free body diagram by force and moment reactions. Due to symmetry, $R = F/2$. We then calculate the bending moment M_b for $0 < x < L/2$ by using the free body diagram between 0 and x:

$$\sum F = 0, \quad R - V = 0, \quad V = R = \frac{F}{2}, \quad \sum M = 0 = M_b - M_R - Vx. \qquad \text{(C.19)}$$

For the beam equation. we have

$$M_b = M_R + \frac{Fx}{2}, \quad \frac{d^2v}{dx^2} = \frac{M_b}{EI}, \quad \frac{d^2v}{dx^2} = \frac{1}{EI}\left(M_R + \frac{Fx}{2}\right). \qquad \text{(C.20)}$$

We integrate the beam equation twice:

$$\frac{dv}{dx} = \frac{1}{EI}\left(M_R x + \frac{Fx^2}{4}\right) + A, \quad v = \frac{1}{EI}\left(M_R\frac{x^2}{2} + \frac{Fx^3}{12}\right) + Ax + B. \qquad \text{(C.21)}$$

As we have three unknowns, A, B and M_R, we apply the following three boundary conditions:

$$v(x = 0) = 0, \quad \left(\frac{dv}{dx}\right)_{x=0}, \quad \left(\frac{dv}{dx}\right)_{x=L/2} = 0. \qquad \text{(C.22)}$$

From the first two it immediately follows that $A = B = 0$, and from the third, which is a symmetry condition, we obtain

$$M_R = -\frac{FL}{8}. \qquad \text{(C.23)}$$

Bringing this result into the deflection equation,

$$v(x) = \frac{Fx^2}{48EI}(-3L + 4x). \qquad \text{(C.24)}$$

Problem 3.5

Using the results found in Problem 3.4, if the load force is $F = 0.1\,\mu N$ located at the middle point as shown in Figure C.5, find the position where the deflection is a maximum and calculate its value. We know that the cantilever has length $L = 1000\,\mu m$, width $W = 10\,\mu m$ and thickness $h = 1\,\mu m$. The cantilever is made of a material having a Young's modulus of $E = 160\,GPa$.

The maximum deflection occurs at $x = L/2$,

$$v_{\text{max}} = v(x = L/2) = -\frac{FL^3}{192EI},$$ (C.25)

the minus sign indicating that the deflection is downwards. The moment of inertia is

$$I = \frac{Wh^3}{12} = 8.33 \times 10^{-25},$$ (C.26)

and the maximum deflection is 3.69 μm.

Problem 3.6

The structure of a crab leg flexure is shown in Figure C.6. It is frequently used in MEMS devices, to support either laterally or vertically moving structures. Calculate the equation for the stiffness along the x-axis.

Following the derivation in [73], we will suppose that a force F_x is applied along the x-axis and that the shin has a fixed end at the bottom and a guided end at the top. We will further require that there is no displacement in the y-direction and that there is no torsion at the end of the thigh of the support. Let us isolate the two parts of the flexure, shin and thigh, as shown

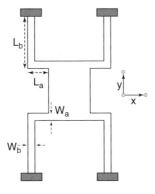

Figure C.6 Problem 3.5

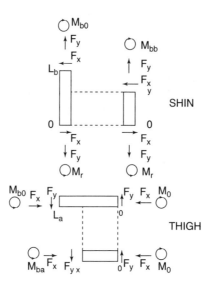

Figure C.7   Free body diagram Problem 3.6

in Figure C.7. We assume a fictitious load F_y and moment M_0. Forces and moments required for equilibrium are shown in Figure C.7.

From the shin and thigh equilibrium conditions we can calculate that the reaction moment M_r, the bending moment M_{b0}, and the two bending moments at the thigh and shin:

$$M_r = M_{b0} - F_x L_b, \quad M_{b0} = M_0 - F_y L_a, \tag{C.27}$$

$$M_{ba} = M_0 - F_y x, \quad M_{bb} = M_0 - F_y L_a - F_x L_b + F_x y. \tag{C.28}$$

The total elastic energy is given by

$$U = U_a + U_b = \int_0^{L_a} \frac{M_{ba}^2}{2EI_a} dx + \int_0^{L_b} \frac{M_{bb}^2}{2EI_b} dy. \tag{C.29}$$

Next, we set the boundary conditions,

$$\frac{\partial U}{\partial M_0} = \theta_0 = 0 \qquad \frac{\partial U}{\partial F_y} = 0. \tag{C.30}$$

From the first equation we get the equation for M_0, and from the second the equation for the fictitious load F_y:

$$M_0 = \frac{F_x L_b^2 + 2F_y L_a L_b + R_{ba} L_a^2 F_y}{2(R_{ba} L_a + L_b)}, \quad F_y = -F_x \frac{3L_b^2}{L_a(R_{ba} L_a + 4L_b)}. \tag{C.31}$$

Finally, the displacement δ_x collinear with the force F_x is given by

$$\delta_x = \frac{\partial U}{\partial F_x}, \tag{C.32}$$

and

$$k_x = \frac{F_x}{\delta_x} = \frac{3EI_b}{L_b^3} \frac{R_{ba}L_a + L_b}{R_{ba}L_a + 4L_b}. \tag{C.33}$$

Problem 3.7

Cantilevers are used commercially for AFM force spectroscopy. In [21] dimensions of commercial cantilevers are collected and can also be found in vendor information. Typical dimensions are length $L = 203.8\ \mu m$, width $W = 20.38\ \mu m$ and thickness $t = 0.55\ \mu m$. Cantilevers are made of silicon nitride. The quoted stiffness constant is 20 pN/nm. Calculate the theoretical value of the stiffness constant assuming a fixed end cantilever.

According to Table 2.5, the Young's modulus of silicon nitride is in the range 166–297 GPa. The cantilever is bent in such a way that the neutral plane is the (x, z) plane, as shown in Figure C.8.

Then the moment of inertia is given by

$$I_z = \frac{Wt^3}{12} = \frac{20.38 \times 10^{-6} \times 0.55^3 \times 10^{-18}}{12} = 2.82 \times 10^{-25}\ m^4. \tag{C.34}$$

The stiffness constant for a cantilever with one end fixed and the other free is given in equation (C.27). Assuming that Young's modulus is $E = 166$ GPa,

$$k = \frac{3EI_z}{L^3} = \frac{3 \times 166 \times 10^9 \times 2.82 \times 10^{-25}}{(203.8 \times 10^{-6})^3} = 0.0166\ \text{N/m}. \tag{C.35}$$

This result is close to the quoted value of $20\,\text{pN/nm} = 0.02\,\text{N/m}$.

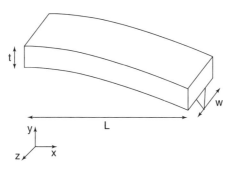

Figure C.8   Problem 3.7

Problem 3.8

For a cantilever of the same dimensions as in Problem 3.7, we would like to calculate the distance at which a CCD camera has to be placed in order to get a displacement of 9 μm of a reflected laser beam at the CCD plane for a force applied of 10 pN.

As can be seen in Figure C.9, the laser beam is reflected vertically when there is no force applied at the tip, but it is reflected at an angle θ when a force is applied.

We know from Section 3.7 that the deflection of a cantilever with an applied force at the tip is given by

$$v = \frac{F}{EI} \left(L\frac{x^2}{2} - \frac{x^3}{6} \right) \tag{C.36}$$

From Figure C.9 it is clear that the angle θ is related to the deflection by

$$\tan\theta = \frac{dv}{dx} = \frac{F}{EI} \left(Lx - \frac{x^2}{2} \right). \tag{C.37}$$

At the tip of the cantilever $x = L$,

$$\tan\theta = \frac{F}{EI} \left(L^2 - \frac{L^2}{2} \right) = \frac{FL^2}{2EI}. \tag{C.38}$$

The angle at the cantilever tip is equal to the reflected beam angle, so

$$\tan\theta = \frac{d}{D} = \frac{FL^2}{2EI}. \tag{C.39}$$

and hence the distance D is

$$D = d\tan\theta = \frac{2EId}{FL^2}. \tag{C.40}$$

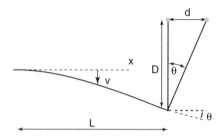

Figure C.9   Problem 3.8

As the moment of inertia for this bending geometry is

$$I_z = \frac{Wt^3}{12},$$ (C.41)

we obtain

$$D = d\tan\theta = \frac{2EWt^3d}{6FL^2}$$ (C.42)

$$= \frac{2 \times 166 \times 10^9 \times 20.38 \times 10^{-6} \times (0.55 \times 10^{-6})^3 \times 9 \times 10^{-6}}{6 \times 10 \times 10^{-9} \times (203.8 \times 10^{-6})^2}$$

$$= 4.06 \times 10^{-3} \text{ m}.$$

Problem 3.9

Find the position of the neutral plane of a cantilever formed by a stack of two films of different materials, as shown in Figure C.10. Calculate the value of b for $t_1 = 1\ \mu m$, $t_2 = 2\ \mu m$, $E_1 = 385$ GPa and $E_2 = 130$ GPa.

Pure bending conditions require that

$$\int_A \sigma_x dA = 0, \quad W \int_{-b}^{a} Eydy = 0, \quad \int_{-b}^{a} Eydy = 0.$$ (C.43)

The latter integral above can be split into two integrals,

$$\int_{-b}^{-b+t_2} E_2 ydy + \int_{-b+t_2}^{+a} E_1 ydy = 0,$$ (C.44)

$$\frac{E_2}{2}((-b+t_2)^2 - b^2) + \frac{E_1}{2}(a^2 - (-b+t_2)^2) = 0.$$ (C.45)

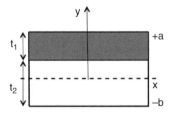

Figure C.10 Problem 3.9

Taking into account that $a + b = t_1 + t_2$,

$$b = \frac{E_2 t_2^2 + E_1 t_1^2 + 2E1t_1 t_2}{2E_2 t_2 + 2E_1 t_1} = 1.895 \text{ μm}, \quad a = 1.105 \text{ μm}. \tag{C.46}$$

Problem 3.10

We know that the differential equation governing the deflection of a beam having an axial tensile stress is the Euler–Bernoulli equation given by

$$EI\frac{d^4 v}{dx^4} - \sigma_0 Wh\frac{d^2 y}{dx^2} = q, \tag{C.47}$$

as given in [15, p. 229] and [16]. Write Matlab code to solve this differential equation with clamped beam boundary conditions at both ends. The beam has a length of $L = 200$ μm, width $W = 10$ μm and thickness $h = 2$ μm. Young's modulus is $E = 130$ GPa. Calculate the value of q required for the beam to deflect 20% of the thickness with stress $\sigma_0 = 0$. Calculate the deflection for the same value of q found in the previous point and for three values of σ_0.

In order to make the solution more general we first proceed to normalize all dimensions in the beam to the beam length L,

$$x' = \frac{x}{L}, \quad y' = \frac{y}{L}. \tag{C.48}$$

Then

$$\frac{dy'}{dx'} = \frac{dy}{dx'} \quad \frac{d^2 y'}{dx'^2} = L\frac{d^2 y}{dx'^2} \quad \frac{d^3 y'}{dx'^3} = L^2 \frac{d^3 y}{dx'^3} \quad \frac{d^4 y'}{dx'^4} = L^3 \frac{d^4 y}{dx^4}. \tag{C.49}$$

This way the variable x' is within the range 0–1. The Euler–Bernoulli equation is written as

$$\frac{d^4 y'}{dx'^4} = \sigma_0 WhL^2 \frac{d^2 y'}{dx'^2} + \frac{qL^3}{EI}. \tag{C.50}$$

For the Matlab simulation we define a vector y as

$$\begin{pmatrix} y(1) \\ y(2) \\ y(3) \\ y(4) \end{pmatrix} = \begin{pmatrix} y' \\ \frac{dy'}{dx'} \\ \frac{d^2 y'}{dx'^2} \\ \frac{d^3 y'}{dx'^3} \end{pmatrix}, \tag{C.51}$$

and the vector *dydx* as

$$
\begin{pmatrix} dydx(1) \\ dydx(2) \\ dydx(3) \\ dydx(4) \end{pmatrix} = \begin{pmatrix} y(2) \\ y(3) \\ y(4) \\ By(3) + A \end{pmatrix},
$$

(C.52)

where

$$
B = \frac{\sigma_0 WhL^2}{EI}, \quad A = \frac{qL^3}{EI}.
$$

(C.53)

```
function euler
%clf
global L sigma0 W h E q
L=200e-6;
sigma0=100e6;
L=200e-6;
W=10e-6;
h=2e-6;
E=130e9; I=(W*(h^3))/12;
q=4e-2;
solinit = bvpinit(linspace(0,1,100),[0 0 0 0 ]);
sol = bvp5c(@twoode,@twobc,solinit);

xint= linspace(0,1);
yint=deval(sol,xint);

function dydx=twoode(x,y)
global L sigma0 W h E q

I=(W*(h^3))/12;
A=q/(E*I)*L^3;
B=12*sigma0/(E*h^2)*L^2;
dydx = [y(2)
        y(3)
        y(4)
        (B*y(3)+A)]
function res=twobc(ya,yb)

global L sigma0 W h E q
res= [ ya(1)
       yb(1)
       ya(2)
       yb(2)]
```

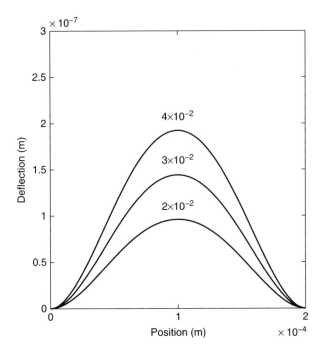

Figure C.11   Solution of Problem 3.10. Plots of deflection along the position in the beam for $q = 2 \times 10^{-2}$, $q = 3 \times 10^{-2}$ and $q = 4 \times 10^{-2}$

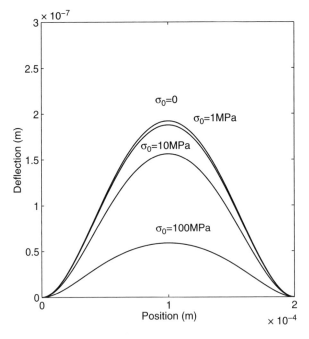

Figure C.12   Results of Problem 3.10. Deflection for $q = 4 \times 10^{-2}$ and several values of σ_0: 0, 1 MPa, 10 MPa and 100 MPa

In Figure C.12, the results are shown for $\sigma_0 = 0$ and several values of q in order to find a suitable value to achieve a maximum deflection value of 20% of the beam thickness. As can be seen, this is achieved for approximately $q = 4 \times 10^{-2}$. For this value of q, we next explore in Figure C.12 the effect of several values of σ_0.

Problem 3.11

Find the definition of the centroid of an area and show that in a rectangular cross-section the neutral line passes through the centroid.

The centroid of an area is defined as a point with the coordinates [74]

$$x_c = \frac{\int_A x \, dA}{\int dA} \tag{C.54}$$

and

$$y_c = \frac{\int_A y \, dA}{\int dA}. \tag{C.55}$$

For a rectangle of side lengths a and b, if we assign the left corner to the origin of coordinates,

$$x_c = \frac{\int_A x \, dA}{\int dA} = \frac{b \int_0^a x \, dx}{ab} = \frac{a}{2} \tag{C.56}$$

and

$$y_c = \frac{a \int_0^b y \, dy}{ab} = \frac{b}{2}. \tag{C.57}$$

When the area has a centre of symmetry the centroid coordinates coincide with the centre of symmetry. The neutral axis in bending is defined by the equation

$$\int_A y \, dA = 0. \tag{C.58}$$

This is true if the origin of coordinates is chosen at the centre of the rectangle as then $dA = a \, dy$:

$$\int_A y \, dA = a \int_{-b/2}^{b/2} y \, dy = 0. \tag{C.59}$$

Problem 3.12

We have a stack of two materials of different thickness and Young's modulus E_1 and E_2, respectively. Show that the stack of two materials can be treated as a single material having two different widths. This is known as the 'equivalent width' method for finding the centroid of the stack.

Consider Figure C.13, where the two models are compared. Material 1 has not changed width W_1, but material 2 has a different width W_2' instead of W_2. Obviously the strain in material 1 is the same in the two cases, and the strain in material 2 should also be the same; then

$$S_2' = S_2, \tag{C.60}$$

and as the two Young's modulus are different,

$$\frac{T_2'}{E_2'} = \frac{T_2}{E_2} = \frac{T_2'}{E_1} \quad \text{and} \quad \frac{T_2'}{T_2} = \frac{E_1}{E_2}, \tag{C.61}$$

as we want $E_2' = E_1$. For the two models to be equivalent, the total force in material 2 should also be the same,

$$T_2'A_2' = T_2A_2 \quad \text{or} \quad \frac{T_2'}{T_2} = \frac{A_2}{A_2'}, \tag{C.62}$$

and then

$$\frac{A_2}{A_2'} = \frac{E_1}{E_2} \tag{C.63}$$

or

$$A_2' = A_2\frac{E_2}{E_1}. \tag{C.64}$$

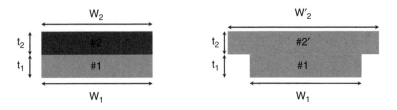

Figure C.13   Problem 3.12

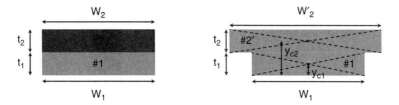

Figure C.14 Problem 3.13

Problem 3.13

Using the equivalent width method, find the centroid of a stack of two layers of different materials with different thickness and Young's modulus.

We first transform the stack into a different sections of the same material as shown in Figure C.13 and then proceed to calculate the centroid coordinates of the two separate rectangles. For each rectangle the centroid is located at its centre of symmetry; then we use the centroid coordinate definitions in Problem 3.11 and shown in Figure C.14:

$$y_c = \frac{\int_A y \, dA}{\int dA} = \frac{y_{c1}A_1 + y_{c2}A_2'}{A_1 + A_2'}. \tag{C.65}$$

From Figure C.14,

$$y_c = \frac{\left(t_1 + \frac{t_2}{2}\right) W_2' t_2 + \frac{t_1}{2} W_1 t_1}{W_1 t_1 + W_2' t_2}. \tag{C.66}$$

D

Chapter 4 Solutions

Problem 4.1

Demonstrate the analytical expression for the rotated piezoresistive coefficient π'_{11}.

The equation we should use is

$$\pi'_{ij} = \sum_{k,l=1}^{6} N_{ik}\pi_{kl}M_{lj};$$

(D.1)

for $i = 1$ and $j = 1$, this becomes

$$\pi'_{11} = \sum_{k,l=1}^{6} N_{1k}\pi_{kl}M_{l1}.$$

(D.2)

We need the components of the matrix (M) that can be found starting from the equation

$$(\rho) = (R)^{-1}(\rho')(R)$$

(D.3)

or

$$\begin{pmatrix} \rho_1 & \rho_6 & \rho_5 \\ \rho_6 & \rho_2 & \rho_4 \\ \rho_5 & \rho_4 & \rho_3 \end{pmatrix} = \begin{pmatrix} l_1 & l_2 & l_3 \\ m_1 & m_2 & m_3 \\ n_1 & n_2 & n_3 \end{pmatrix} \cdot \begin{pmatrix} \rho'_1 & \rho'_6 & \rho'_5 \\ \rho'_6 & \rho'_2 & \rho'_4 \\ \rho'_5 & \rho'_4 & \rho'_3 \end{pmatrix} \cdot \begin{pmatrix} l_1 & m_1 & n_1 \\ l_2 & m_2 & n_2 \\ l_3 & m_3 & n_3 \end{pmatrix}.$$

(D.4)

Multiplying the $(R)^{-1}$ and (ρ') matrices, equation (D.4) becomes

$$\begin{pmatrix} \rho_1 & \rho_6 & \rho_5 \\ \rho_6 & \rho_2 & \rho_4 \\ \rho_5 & \rho_4 & \rho_3 \end{pmatrix} = \begin{pmatrix} l_1\rho'_1 + l_2\rho'_6 + l_3\rho'_5 & l_1\rho'_6 + l_2\rho'_2 + l_3\rho'_4 & l_1\rho'_5 + l_2\rho'_4 + l_3\rho'_3 \\ m_1\rho'_1 + m_2\rho'_6 + m_3\rho'_5 & m_1\rho'_6 + m_2\rho'_2 + m_3\rho'_4 & m_1\rho'_5 + m_2\rho'_4 + m_3\rho'_3 \\ n_1\rho'_1 + n_2\rho'_6 + n_3\rho'_5 & n_1\rho'_6 + n_2\rho'_2 + n_3\rho'_4 & n_1\rho'_5 + n_2\rho'_4 + n_3\rho'_3 \end{pmatrix}$$

$$\cdot \begin{pmatrix} l_1 & m_1 & n_1 \\ l_2 & m_2 & n_2 \\ l_3 & m_3 & n_3 \end{pmatrix}.$$

(D.5)

Understanding MEMS: Principles and Applications, First Edition. Luis Castañer.
© 2016 John Wiley & Sons, Ltd. Published 2016 by John Wiley & Sons, Ltd.
Companion Website: www.wiley.com/go/castaner/understandingmems

So

$$\rho_1 = \left(l_1\rho_1' + l_2\rho_6' + l_3\rho_5'\right)l_1 + \left(l_1\rho_6' + l_2\rho_2' + l_3\rho_4'\right)l_2 + \left(l_1\rho_5' + l_2\rho_4' + l_3\rho_3'\right)l_3, \quad (D.6)$$

which is readily reorganized as

$$\rho_1 = \rho_1'l_1^2 + \rho_2'l_2^2 + \rho_3'l_3^2 + \rho_4'(2l_2l_3) + \rho_5'(2l_3l_1) + \rho_6'(2l_1l_2). \quad (D.7)$$

Similar calculations for the other components lead to

$$(M) = \begin{pmatrix} l_1^2 & l_2^2 & l_3^2 & 2l_3l_2 & 2l_3l_1 & 2l_2l_1 \\ m_1^2 & m_2^2 & m_3^2 & 2m_3m_2 & 2m_3m_1 & 2m_2m_1 \\ n_1^2 & n_2^2 & n_3^2 & 2n_3n_2 & 2n_3n_1 & 2n_2n_1 \\ m_1n_1 & m_2n_2 & m_3n_3 & m_2n_3+m_3n_2 & m_3n_1+m_1n_3 & m_1n_2+m_2n_1 \\ n_1l_1 & n_2l_2 & n_3l_3 & n_2l_3+n_3l_2 & n_3l_1+n_1l_3 & n_1l_2+n_2l_1 \\ l_1m_1 & l_2m_2 & l_3m_3 & m_2l_3+m_3l_2 & m_3l_1+m_1l_3 & m_1l_2+m_2l_1 \end{pmatrix}. \quad (D.8)$$

Going back to equation (D.2) and taking into account that a large number of elements of the (Π) matrix are zero, we can write

$$\pi_{11}' = \sum_{k,l=1}^{6} N_{1k}\pi_{kl}M_{l1} \quad (D.9)$$

$$= N_{11}\pi_{11}M_{11} + N_{11}\pi_{12}M_{21} + N_{11}\pi_{12}M_{31} + N_{12}\pi_{12}M_{11} + N_{12}\pi_{11}M_{21}$$

$$+N_{12}\pi_{12}M_{31} + N_{13}\pi_{12}M_{11} + N_{13}\pi_{12}M_{21} + N_{13}\pi_{44}M_{41} + N_{15}\pi_{44}M_{51}$$

$$+N_{16}\pi_{44}M_{61}.$$

Substituting and grouping,

$$\pi_{11}' = \pi_{11}\left(l_1^4 + m_1^4 + n_1^4\right) + 2(\pi_{12} + \pi_{44})\left(m_1^2l_1^2 + n_1^2l_1^2 + m_1^2n_1^2\right). \quad (D.10)$$

Taking into account that l_1, m_1 and n_1 satisfy the equation

$$l_1^2 + m_1^2 + n_1^2 = 1, \quad (D.11)$$

equation (D.10) can be written

$$\pi_{11}' = \pi_{11}\left(l_1^2\left(1 - m_1^2 - n_1^2\right) + m_1^2\left(1 - l_1^2 - n_1^2\right) + n_1^2\left(1 - l_1^2 - m_1^2\right)\right) \quad (D.12)$$

$$+2(\pi_{12} + \pi_{44})\left(m_1^2l_1^2 + n_1^2l_1^2 + m_1^2n_1^2\right)$$

Grouping again,

$$\pi_{11}' = \pi_{11}\left(l_1^2 + m_1^2 + n_1^2\right) + 2(\pi_{12} + \pi_{44} - \pi_{11})\left(m_1^2l_1^2 + n_1^2l_1^2 + m_1^2n_1^2\right). \quad (D.13)$$

Taking into account equation (D.11), we finally obtain

$$\pi'_{11} = \pi_{11} - 2\pi_0 \left(l_1^2 m_1^2 + l_1^2 n_1^2 + m_1^2 n_1^2 \right),$$ (D.14)

with

$$\pi_0 = \pi_{11} - \pi_{12} - \pi_{44}.$$ (D.15)

Problem 4.2

The National Highway Traffic Safety Administration (NHTSA) recommends a deceleration of 60g to deploy airbags, although the driver's weight and other factors are also taken into consideration by the electronic trigger system. In this problem we would like to evaluate the performance of an accelerometer made of a cantilever supporting an inertial mass at the tip and a piezoresistor located at the cantilever end closest to the support.

Assuming that the inertial mass is a silicon volume of 2.5×10^{-12} kg, calculate the force when the acceleration is 100g. Considering this as a point force applied to the tip of the cantilever itself, calculate the deflection of the tip as a function of the length L and the value for $L = 1000\,\mu m$. The cantilever is made of silicon, and has width $W = 50\,\mu m$ and thickness $t = 2\,\mu m$. Assume that the edge of the cantilever is clamped. Young's modulus for silicon is $E = 164\,GPa$.

The silicon density is 2339 kg/m^3, so

$$F = V \times d \times 100 \times g,$$ (D.16)

$$F = 2.5 \times 10^{-12} \times 100 \times 9.8 = 2.45 \times 10^{-9} \text{N}.$$ (D.17)

For $L = 1000\,\mu m$,

$$I = \frac{Wt^3}{12} = 3.3 \times 10^{-23} \quad \text{and} \quad v(L) = \frac{F}{EI}\left(\frac{L^3}{2} - \frac{L^3}{6}\right) = \frac{FL^3}{3EI} = 0.15\mu. \quad \text{(D.18)}$$

Problem 4.3

For the same data as in Problem 4.2, calculate the distribution of stress along the cantilever length as a function of the position y inside the cantilever above the neutral surface. Calculate the value of the maximum stress and identify the location where this occurs.

According to equation (3.4),

$$\sigma_x = -E\frac{y}{\rho_c}, \quad \frac{1}{\rho_c} = \frac{M_b}{EI}, \quad M_b = F(L - x). \tag{D.19}$$

It follows that

$$\sigma = \frac{12F(L - x)}{Wt^3}y. \tag{D.20}$$

The maximum occurs at $x = 0$ and $y = \pm t/2$, and hence

$$\sigma = \frac{6F(L)}{Wt^2}. \tag{D.21}$$

Problem 4.4

For the same data as in Problem 4.2, if a piezoresistor is placed where the maximum stress occurs, calculate the value of the relative change in the resistance when the device is subject to 100 g acceleration. Plot the values of the deflection, stress and $\Delta R/R$ as a function of the cantilever length L.

We have

$$\frac{\Delta R}{R} = \Pi_l\sigma = \frac{\pi_{44}}{2}\sigma. \tag{D.22}$$

The numerical results are shown in Figure D.1 as a function of the length of the cantilever L. As can be seen, for a cantilever length of $1000\,\mu m$, the deflection is close to $100\,\mu m$, the stress close to 70 MPa and the relative change of the resistance approximately 3%.

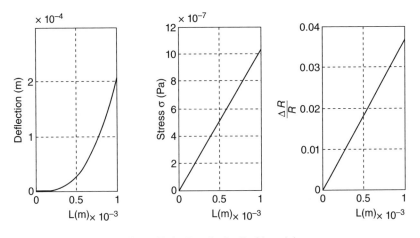

Figure D.1   Results for Problem 4.4

Problem 4.5

Calculate the output of a Wheatstone bridge such as the one shown in Example 4.6.1 of piezoresistances made of polysilicon film, located as shown in Figure D.2 on top of a membrane clamped in the periphery.

We will use the results shown in Figure 4.7 for the stresses in the membrane, where plots of the normalized values of the stress are shown. The plots displayed correspond to $\sigma_x h^2/(p_0 L^2)$; h is the membrane thickness, L is the side length and P is the applied pressure. From Figure 4.7 identify the corresponding values of the longitudinal and transversal stresses for the four resistors. For polysilicon resistors the piezoresistive coefficients are: longitudinal, $\pi'_l = 0.425\pi_{44}$; transversal, $\pi'_t = -0.2\pi_{44}$. The shear stress can be neglected. Also $\pi_{44} = 138 \times 10^{-11}$ Pa^{-1}.

Due to the location of the resistors, the corresponding values of the longitudinal and transversal resistors are found at the following coordinates in the plot: for R_1 and R_4,

$$\sigma_l = \sigma_x(0.5, 0) \simeq 0.294 \frac{p_0 L^2}{h^2}, \quad \sigma_t = \sigma_x(0, 0.5) \simeq 0.115 \frac{p_0 L^2}{h^2}; \tag{D.23}$$

for R_2 and R_3,

$$\sigma_l = \sigma_t = \sigma_x(0, 0) \simeq -0.13 \frac{p_0 L^2}{h^2}. \tag{D.24}$$

Then

$$\frac{\Delta R_1}{R_1} = \frac{\Delta R_4}{R_4} = \pi'_l 0.294 \frac{p_0 L^2}{h^2} + 0.115 \pi'_t \frac{p_0 L^2}{h^2}, \tag{D.25}$$

$$\frac{\Delta R_2}{R_2} = \frac{\Delta R_3}{R_3} = -\pi_{44} 0.13 \frac{p_0 L^2}{h^2} - 0.13 \pi'_t \frac{p_0 L^2}{h^2}. \tag{D.26}$$

Using the parameter values,

$$\frac{\Delta R_1}{R_1} = \frac{\Delta R_4}{R_4} = 0.101 \pi_{44} \frac{p_0 L^2}{h^2}, \tag{D.27}$$

$$\frac{\Delta R_2}{R_2} = \frac{\Delta R_3}{R_3} = -0.029 \pi_{44} \frac{p_0 L^2}{h^2}. \tag{D.28}$$

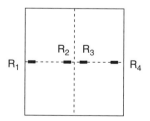

Figure D.2 Problem 4.5

Problem 4.6

We place four resistors in a Wheatstone bridge as in Example 4.6.1. Write the equation giving the output voltage as a function of the relative increment of the resistors after applying a pressure difference of P. Calculate the value of the output voltage when $P = 1000$ Pa, $\ell = 500\ \mu m$, $h = 1\ \mu m$, and $\pi_{44} = 138 \times 10^{-11}$.

Solving the circuit equations for the Wheatstone bridge,

$$V_0 = V_{CC} \frac{1 + \frac{\Delta R_3}{R_3}}{2 + \frac{\Delta R_1}{R_1} + \frac{\Delta R_3}{R_3}} - V_{CC} \frac{1 + \frac{\Delta R_4}{R_4}}{2 + \frac{\Delta R_2}{R_2} + \frac{\Delta R_4}{R_4}}. \tag{D.29}$$

Substituting the data values,

$$V_0 = -0.33 \text{ V}. \tag{D.30}$$

Problem 4.7

We have a pressure sensor based on a circular membrane of radius a and we would like to compare the sensitivity provided by a piezoresistor oriented radially at the edge of the membrane (Figure D.3(a)) with a piezoresistor oriented tangentially also at the edge of the membrane (Figure D.3(b)).

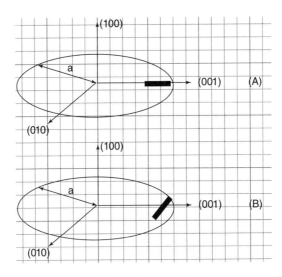

Figure D.3 Problem 4.7

The membrane is made out of n-type silicon (100) having longitudinal piezoresistance coefficient $\pi_l = -102,2 \times 10^{-11}\ Pa^{-1}$, and transversal piezoresistance coefficient $\pi_t = 53.4 \times 10^{-11}$ Pa^{-1}. These values correspond to the crystallographic coordinates, $v = 0.33$, $a = 500\ \mu m$ and $h = 1\ \mu m$.

Calculate the relative change of resistance when the differential pressure P applied is 10^4 Pa for the two cases. What position of the piezoresistance gives better sensitivity?

There is no need for rotation of coordinates as the application axes are the same as the crystallographic axes. We have

$$\sigma_r = \frac{3a^2}{4h^2}P = 1.87\,\text{GPa}, \quad \sigma_t = v\frac{3a^2}{4h^2}P = 0.618\,\text{GPa}, \tag{D.31}$$

and

$$\left(\frac{\Delta R}{R}\right)_r = \pi_l\sigma_l = -1.91, \quad \left(\frac{\Delta R}{R}\right)_t = \pi_t\sigma_t = 0.33. \tag{D.32}$$

Thus the radial geometry gives better sensitivity.

Problem 4.8

We have a clamped circular membrane of thickness $h = 2\ \mu m$ and radius $a = 200\ \mu m$. The Poisson ratio is $v = 0.27$. For a pressure difference of $P = 1000$ Pa, calculate the maximum of the radial stress and the maximum of the absolute value of the tangential stress. Let us put a piezoresistor along the radial direction at the point of maximum radial stress. We know that this direction is (011). Using the results of the piezoresistive coefficients for these rotated axes from Example 4.5.1, calculate the value of the relative change of the piezoresistor ($\Delta R/R$).

From the equation

$$\sigma_r = -\frac{3P_0a^2}{8h^2}\left(3\frac{r^2}{a^2} - 1 + v\frac{r^2}{a^2} - v\right), \tag{D.33}$$

we can see that the maximum is located at $r = a$, thus

$$\sigma_r = -\frac{3P_0a^2}{4h^2} = 0.037\,\text{MPa}. \tag{D.34}$$

The tangential stress at $r = 0$ and $r = a$ is given by

$$\sigma_t(r = a) = -\frac{3P_0a^2}{4h^2}v, \quad \sigma_t(r = 0) = -\frac{3P_0a^2}{8h^2}(1 + v). \tag{D.35}$$

The maximum, at the membrane centre, $r = 0$, is $\sigma_{t\max} = 0.018$ MPa. For comparison, in a square membrane, the stress normal to the edge at the centre of the side of the membrane is $\sigma_n = 5.88$ MPa, and the stress parallel to the edge at the centre of the side of the membrane is $\sigma_p = 3$ MPa. Finally,

$$\frac{\Delta R}{R} = \pi'_{11}\sigma_r + \pi'_{12}\sigma_t = 1.44 \times 10^{-5}. \tag{D.36}$$

Problem 4.9

We have a bimorph made of a stack of two materials, one of them piezoelectric. If we bias the piezoelectric material with a voltage V, the electric field created (assumed to be in direction 3) is V/d, where d is the thickness of the piezoelectric material, which is a fraction of the total thickness of the bimorph. Calculate the equation giving the stress in the direction of axis 1. Calculate the bending moment corresponding to this stress and solve the beam equation to find the deflection as a function of the position and the deflection at the tip.

The strain generated along axis 1 is

$$S_1 = d_{31}\frac{V}{d}, \tag{D.37}$$

and the corresponding stress

$$T_1 = d_{31}E\frac{V}{d}, \tag{D.38}$$

where E is Young's modulus.

If we only have stress in the piezoelectric layer, the moment will be

$$M_b = -\int_A T_1 y dA, \tag{D.39}$$

where y is the distance of the point to the neutral axis. As T_1 is different from zero only in the piezoelectric layer, the moment is

$$M_b = -W\int_0^d d_{31}E\frac{V}{d}y dy = -Wd_{31}E\frac{V}{d}\frac{d^2}{2} = -Wd_{31}E\frac{Vd}{2}. \tag{D.40}$$

As can be seen, this moment is not a function of the position x [75], so the beam equation is

$$\frac{d^2v}{dx^2} = \frac{M_b}{EI} = -Wd_{31}\frac{Vd}{2I}. \tag{D.41}$$

Integrating twice with fixed-end boundary conditions at $x = 0$,

$$v(x) = -Wd_{31}\frac{Vd}{4I}x^2. \tag{D.42}$$

The maximum deflection will occur at the tip $x = L$, so

$$v_{\text{max}} = v(x = L) = -Wd_{31}\frac{Vd}{4I}L^2, \tag{D.43}$$

and

$$v(x) = v_{\text{max}}\left(\frac{x}{L}\right)^2, \tag{D.44}$$

which is the typical parabolic deflection often encountered in piezoelectric structures.

E

Chapter 5 Solutions

Problem 5.1

For the same data as in Example 5.3.1, plot the values of g as a function of the applied voltage.

The steady-state equilibrium equation is equation (5.19),

$$\frac{V^2}{2} \frac{\epsilon_0 A}{g^2} = k(g_0 - g), \tag{E.1}$$

which we rearrange as

$$2kg^3 - 2kg_0g^2 + \epsilon_0 AV^2 = 0 \tag{E.2}$$

This is a third-degree equation that has three solutions. As seen in Example 5.3.1, there are only two equilibrium points and the third solution gives a negative value and is discarded. From the two other solutions, the only stable equilibrium solution is that greater than $\frac{2}{3}g_0$. We plot in Figure E.1 the results of the value of the stable equilibrium point for each value of the voltage.

Problem 5.2

Calculate the total energy consumed from the source during the switching of an actuator of the same parameter values as in Example 5.3.1.

The total energy drawn from the source to drive the MEMS actuator splits into several components: kinetic energy due to the movement of the movable plate, elastic energy stored in the spring, energy stored in the capacitor, dissipation in the resistor and dissipation in the

Understanding MEMS: Principles and Applications, First Edition. Luis Castañer.
© 2016 John Wiley & Sons, Ltd. Published 2016 by John Wiley & Sons, Ltd.
Companion Website: www.wiley.com/go/castaner/understandingmems

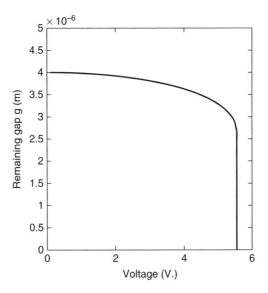

Figure E.1   Problem 5.1

damper. The equations giving the instantaneous values of these energy components are [33]:

$$U_k = \frac{1}{2}mv^2 \qquad \text{kinetic energy}$$

$$U_{el} = \frac{1}{2}kx^2 \qquad \text{elastic energy}$$

$$U_{cap} = \frac{1}{2}CV^2 = \frac{q^2}{2C} \qquad \text{capacitor energy}$$

$$U_R = \int_0^t \left(\frac{dq}{dt}\right)^2 Rdt \qquad \text{resistor energy}$$

$$U_{damp} = \int_0^t bv^2 dt \qquad \text{damping energy}$$

$$U_{source} = V_s q \qquad \text{source energy}$$

where v is the velocity dx/dt. All the energy is provided by the source and hence:

$$U_{source} = U_k + U_{el} + U_{cap} + U_R + U_{damp}$$

Solving the equations we find the three state variables x, v and q, and from them we calculate the energy from the source and the elastic, kinetic and capacitor energies. Subtracting the last three from the source energy, we get the total damping and resistor losses. From the results in Figure E.2 it can be seen that although we have restricted the calculation to a time span not very

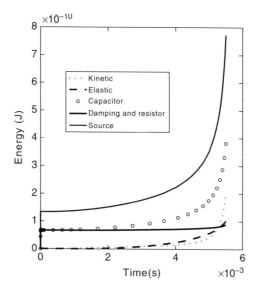

Figure E.2   Problem 5.2

close to the pull-in, the total energy drawn from the source and particularly the components of the capacitor energy and kinetic energy increase sharply (see [33] for more details of the energy distribution during the transient).

Problem 5.3

One way to extend the travel range of pull-in-based actuators is to add an external capacitor C_{ext} in series with the actuator as shown in Figure E.3. Show that the condition required to get the full stable range is that $C_0/C_{ext} > 2$, where $C_0 = \epsilon_0 A/g_0$. Calculate the equation giving the voltage V that should be applied to get full gap stability.

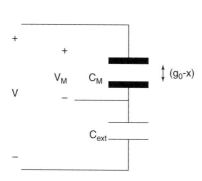

Figure E.3   Problem 5.3

The electrostatic force is related to the capacitance of the intrinsic device C_M by

$$F_E = \frac{Q_M^2}{2\epsilon_0 A} = \frac{C_M V_M^2}{2\epsilon_0 A}. \tag{E.3}$$

Taking into account that

$$C_M = \frac{2\epsilon_0 A}{g_0 - x}, \tag{E.4}$$

the force is given by

$$F_E = \frac{\epsilon_0 A V_M^2}{2(g_0 - x)^2}. \tag{E.5}$$

In equilibrium the mechanical ($F_M = kx$) and electrostatic forces are equal, hence

$$\frac{\epsilon_0 A V_M^2}{2(g_0 - x)^2} = kx. \tag{E.6}$$

The stability condition is

$$\frac{d}{dx}(F_E - F_M) < 0, \tag{E.7}$$

which can be written as

$$\frac{d}{dx}\left(\frac{\epsilon_0 A V_M^2}{2(g_0 - x)^2} - kx\right) < 0. \tag{E.8}$$

Calculating the derivative,

$$\frac{\epsilon_0 A}{2}\left(\frac{2V_M \frac{dV_M}{dx}(g_0 - x)^2) + 2V_M^2(g_0 - x)}{(g_0 - x)^4}\right) - k < 0. \tag{E.9}$$

Taking into account that from equation (E.6),

$$\frac{\epsilon_0 A}{2} = \frac{kx(g_0 - x)^2}{V_M^2}, \tag{E.10}$$

the stability condition becomes

$$\frac{2V_M x}{V_M^2}\frac{dV_M}{dx} + \frac{2x}{g_0 - x} - 1 < 0. \tag{E.11}$$

In this circuit we have two capacitors C_M and C_{ext} in series. Then

$$C_M V_M = C_{ext} V_S, \tag{E.12}$$

$$V = V_M + V_S. \tag{E.13}$$

The voltage V_M can then be written as

$$V_M = V \frac{C_{ext}}{C_{ext} + C_M} \tag{E.14}$$

and

$$V_M = V \frac{C_{ext}(g_0 - x)}{C_{ext}(g_0 - x) + \epsilon_0 A}. \tag{E.15}$$

In order to find the stability condition we need the derivative of V_M,

$$\frac{dV_M}{dx} = V C_{ext} \frac{-\epsilon_0 A}{(C_{ext}(g_0 - x) + \epsilon_0 A)^2}. \tag{E.16}$$

Substituting into equation (E.11),

$$\frac{-2x\epsilon_0 A}{C_{ext}(g_0 - x) + \epsilon_0 A} + 2x - (g_0 - x) < 0. \tag{E.17}$$

Solving for x,

$$x < \frac{g_0}{3} \left(1 + \frac{\epsilon_0 A}{g_0 C_{ext}} \right). \tag{E.18}$$

Defining

$$C_0 = \frac{\epsilon_0 A}{g_0}, \tag{E.19}$$

the stability condition is given by

$$x < \frac{g_0}{3} \left(1 + \frac{C_0}{C_{ext}} \right). \tag{E.20}$$

The condition for full gap stability is found by letting $x = g_0$ in equation (E.20), leading to

$$\frac{C_0}{C_{ext}} > 2. \tag{E.21}$$

This means that full gap stability can be reached provided that the initial capacitance of the device is at least twice larger than the external capacitor. It is important to know whether adding this external capacitor modifies the driving conditions. For that, we can calculate the value of the source voltage V required for full gap. From the equilibrium condition in equation (E.10) and the voltage equation (E.14), we can write

$$\frac{\epsilon_0 A V^2 C_{\text{ext}}^2}{2\left(C_{\text{ext}}(g_0 - x) + \epsilon_0 A\right)^2} = kx. \tag{E.22}$$

For full gap at $x = g_0$,

$$\frac{V^2 C_{\text{ext}}^2}{2\epsilon_0 A} = kg_0. \tag{E.23}$$

Thus, we finally obtain that the voltage required is given by

$$V = \frac{1}{C_{\text{ext}}} \sqrt{2\epsilon_0 A k g_0}. \tag{E.24}$$

Problem 5.4

For the same circuit as in Problem 5.3, we have a MEMS switch of area $200 \times 200 \, \mu m^2$, $g_0 = 1 \, \mu m$, $k = 5.2 \, N/m$. (1) Calculate the required value of the external capacitance C_{ext}. (2) Calculate the value of the voltage required and compare it with the pull-in voltage of the MEMS switch.

We have

$$C_0 = \frac{\epsilon_0 A}{g_0} = 0.35 \, \text{pF}, \quad C_{\text{ext}} = 0.175 \, \text{pF}.$$

Now

$$V = \frac{1}{C_{\text{ext}}} \sqrt{2\epsilon_0 A k g_0} = 10.9 \, \text{V}, \quad V_{\text{PI}} = \sqrt{\frac{8kg_0^3}{27\epsilon_0 A}} = 2.08 \, \text{V},$$

so

$$\frac{V}{V_{\text{PI}}} = 5.19.$$

Problem 5.5

A MEMS parallel-plate actuator voltage driven is driven beyond pull-in. To avoid short-circuiting the voltage source, a thin dielectric is deposited on top of one of the electrodes. (1) Calculate the value of the capacitance when the switch is open. (2) Calculate the capacitance when the switch is closed. (3) Calculate the tunable range $C_{\text{ON}}/C_{\text{OFF}}$. Take $g_0 = 2 \, \mu m$, $t = 200 \, nm$, $\epsilon_r = 3.9$ and $\epsilon_0 = 8.85 \times 10^{-12} \, F/m$.

The model we use is that for two series capacitors, the gap capacitor C_g, and the dielectric capacitor C_d. When the voltage applied is zero, we have

$$\frac{1}{C_{OFF}} = \frac{1}{C_g} + \frac{1}{C_d} = \frac{g_0}{\epsilon_0 A} + \frac{t}{\epsilon_0 \epsilon_r A}, \quad C_{OFF} = \frac{\epsilon_0 A}{g_0 + \frac{t}{\epsilon_r}}.$$

When the voltage is larger than the pull-in voltage, we have

$$C_{ON} = \frac{\epsilon_0 \epsilon_r A}{t}, \quad \frac{C_{ON}}{C_{OFF}} = \frac{\epsilon_r}{t}\left(g_0 + \frac{t}{\epsilon_r}\right) = 1 + \frac{\epsilon_r g_0}{t} = 37.$$

Problem 5.6

A linear electrostatic motor is based on a distribution of electrodes such as shown in Figure E.4. (1) Indicate the sign of the movement when a voltage is applied between electrode 1 and the upper electrode. (2) Indicate the sign of the movement when the voltage is applied between electrode 3 and the upper electrode. (3) Calculate the strength of the force in both cases. Take $g_0 = 2\,\mu m$, $\epsilon_0 = 8.85 \times 10^{-12}$ F/m, $W = 20\,\mu m$ and $V = 10$ V.

(1) The movement is towards electrode 1. (2) The movement is towards electrode 3. The force is the same in both cases:

$$F = \frac{\epsilon_0 V^2 W}{2 g_0} = 4.42\,\text{N}.$$

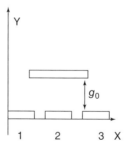

Figure E.4   Problem 5.6

Problem 5.7

A parallel-plate MEMS switch is used to implement a voltage DC/DC step-up converter. The switch is first driven beyond pull-in, the upper plate collapses and the capacitance is charged. A short-circuit is avoided by depositing a thin dielectric film on top of the bottom plate. Then a mechanical pull force is applied to raise the plate to the up position. The diode shown

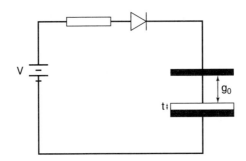

Figure E.5   Problem 5.7

in Figure E.5 blocks the discharge of the capacitor in this phase. Calculate the amount of charge received by the capacitor in the charging phase. Calculate the value of the voltage across the capacitor after the pulling phase has ended. Take $g_0 = 2 \, \mu m$, $t = 100 \, nm$, $\epsilon_r = 3.9$, $\epsilon_0 = 8.85 \times 10^{-12}$ F/m and area $200 \times 200 \, \mu m^2$.

If the charging time constant is short and the source V is connected until the plates collapse, the charge delivered to the MEMS switch is

$$Q = C_{ON}V = \frac{\epsilon_0 \epsilon_r A}{t} V = 1.38 \times 10^{-11} F \times 5V = 6.9 \times 10^{-11} \text{ C.}$$

The charge remains in the MEMS switch and when the plates are lifted apart

$$Q = C_{OFF}V' \quad \text{and} \quad V' = \frac{Q}{C_{OFF}} = \frac{Q}{\frac{\epsilon_0 A}{g_0 + \frac{t}{\epsilon_r}}} = \frac{6.9 \times 10^{-11}}{1.74 \times 10^{-13}} = 396.5 \text{ V.}$$

Problem 5.8

Solve the differential equation for the dynamic response of the parallel-plate actuator if a delta pulse of charge is applied at $t = 0$ with a value of $Q_s = 8 \times 10^{11}$ C, and find the time variation of the deflection for $t > 0$. The data values are: $b = 0.17$, $m = 15.4 \times 10^{-6}$, $k = 4.6$, $g_0 = 8 \times 10^{-6}$, $A = 1.2 \times 10^{-6}$, $\epsilon_0 = 8.85 \times 10^{-12}$ and $Q_s = 8 \times 10^{-11}$. Recall that $C_0 = \epsilon_0 A / g_0$.

If a charged delta pulse is applied at $t = 0$, the movable plate does not have time to move before all the charge has been injected at the capacitor, thus the mass–spring–damper model is given by:

$$m\frac{d^2x}{dt^2} + b\frac{dx}{dt} + kx = \frac{Q^2}{2C_0 g_0}.$$

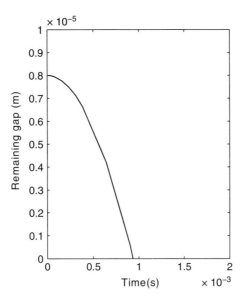

Figure E.6   Results of Problem 5.8

Now we only have two state variables, x and v, as the charge is a constant Q_S. The resulting equation has an analytical solution but we can still use Matlab code to solve the problem. We formulate the equation above as follows:

$$\frac{dx}{dt} = v$$

$$\frac{dv}{dt} = -2\omega_0\zeta v - \omega_0^2 x + \frac{Q_S^2}{2C_0 g_0 m}$$

where $\omega_0 = \sqrt{k/m}$ and $\zeta = b/2\sqrt{km}$. The results are shown in Figure E.6.

Problem 5.9

We would like to use a torsional actuator, such as the one depicted in Figure 5.12, to measure angle by measuring capacitance. Calculate the maximum tilt angle we can have. Calculate the change in capacitance when the change in angle is from $\theta_{max}/2$ to $\theta_{max}/3$. Take $L = 15\,\mu m$, $b = 12\,\mu m$, $d = 5\,\mu m$.

We have

$$\sin\theta_{max} = \frac{d}{L} = \frac{1}{3}, \quad \text{so } \theta_{max} = 19.47°.$$

From equation (5.65),

$$dC = \frac{\epsilon_0 A}{y} = \frac{\epsilon_0 W dx}{y},$$

where W is the width of the plate. From Figure 5.14 we have $\tan \theta = (y - d)/x$, and hence

$$dC = \frac{\epsilon_0 W dx}{d + x \tan \theta}.$$

We change variables

$$z = d + x \tan \theta,$$

yielding

$$dz = \tan \theta dx$$

and

$$dC = \frac{\epsilon_0 W}{z \tan \theta} dz.$$

Integrating yields

$$C = \int_{z_1}^{z_2} \frac{\epsilon_0 W}{\tan \theta} \frac{dz}{z} = \frac{\epsilon_0 W}{\tan \theta} \ln \frac{z_2}{z_1}.$$

As $z_2 = d + x_2 \tan \theta$, $z_1 = d + x_1 \tan \theta$ and $x_2 = x_1 + b$ with $x_1 \to \infty$,

$$C \simeq \frac{\epsilon_0 W}{\tan \theta} \ln \left(1 + \frac{b \tan \theta}{d} \right).$$

With the data given,

$$C \simeq \frac{1.06 \times 10^{-16}}{\tan \theta} \ln \left(1 + \frac{12 \tan \theta}{5} \right),$$

and

$$C(\theta_{\max}/2) = 2.12 \times 10^{-16} \, \text{F}, \quad C(\theta_{\max}/3) = 2.23 \times 10^{-16} \, \text{F}.$$

The change in capacitance is 4.9%.

Problem 5.10

In an electrostatic actuator of two parallel plates moving against each other, we want to know the condition required to have just one equilibrium point. The plates have a surface of $A = 200\,\mu m \times 200\,\mu m$, the air permittivity is 8.85×10^{-12} F/m, the initial gap is $g_0 = 2\,\mu m$ and $k = 1$ N/m. Write the condition required in order to have a single equilibrium point. Calculate the value of the gap g satisfying this requirement. Is this point stable? Calculate the voltage required to reach this equilibrium point.

We know that equations of the electrostatic force F_E and the mechanical restoring force F_M as functions of the gap g:

$$F_E = \frac{\epsilon_0 A V^2}{2g^2} \quad \text{and} \quad F_M = k(g_0 - g). \tag{E.25}$$

A single equilibrium point exists when the plot of the electrostatic force and the plot of the mechanical force are tangent to each other, hence the condition is

$$\frac{dF_E}{dg} = \frac{dF_M}{dg}. \tag{E.26}$$

We have

$$\frac{dF_M}{dg} = -\frac{\epsilon_0 A V^2}{g^3} \quad \text{and} \quad \frac{dF_m}{dg} = -k, \tag{E.27}$$

hence

$$-\frac{\epsilon_0 A V^2}{g^3} = -k. \tag{E.28}$$

From the equilibrium condition,

$$F_E = \frac{\epsilon_0 A V^2}{2g^2} = F_M = k(g_0 - g). \tag{E.29}$$

Substituting k from equation (E.28) into equation (E.29),

$$g = \frac{2}{3}g_0 = 1.33\,\mu m. \tag{E.30}$$

The equilibrium point is at the frontier between stability and instability and hence corresponds to the pull-in condition, The required voltage is precisely the pull-in voltage,

$$V_{\text{PI}} = \sqrt{\frac{2k(g_0 - g)g^2}{\epsilon_0 A}} = 2.587\ \text{V}. \tag{E.31}$$

Problem 5.11

We have two ways of implementing a movement to close a gap as shown in Figure E.7, torsional actuation or a comb actuator, and we are interested in calculating the voltage required in the two cases and also the total energy drawn from the source. Take $y_0 = 1\ \mu m$, $t = 1\ \mu m$, $k = 1\ N/m$, $\epsilon_0 = 8.85 \times 10^{-12}\ F/m$, $g_0 = 1\ \mu m$, $d = 2\ \mu m$, $L = 5\ \mu m$, $W = 12\ \mu m$, $k_\theta = 8 \times 10^{-14}\ N/mrad$. We use a circuit to drive the actuator, composed of a DC voltage source V_{CC} having the same value as the voltage required to close the gap and a series resistor $R = 1\ k\Omega$.

For the comb actuator, the upper plate is moving downwards as shown by the arrow in Figure E.7. The thickness of the comb is t and the comb actuator gap is g_0. We have

$$F_E = \frac{2\epsilon_0 t V^2}{g_0} \quad \text{and} \quad F_M = k(y - y_0). \tag{E.32}$$

Setting $F_E = F_M$, we find $V = 97.03$ V, which is the voltage required to close the gap of 2 μm. The total energy drawn from the source of a voltage $V_{CC} = V$ is given by

$$E = \int V_{CC} i\, dt = V_{CC} \int_0^Q dQ = V_{CC} Q. \tag{E.33}$$

As the voltage applied is larger than the pull-in voltage, the superimposed area between fingers is

$$C_{\max} = 4\frac{\epsilon_0 t y}{g_0} = 1.27 \times 10^{-15}\ \text{F}, \tag{E.34}$$

and hence the energy from the source is

$$E = V_{CC} Q = V_{CC}^2 C_{\max} = 12.02 \times 10^{-12}\ \text{J}. \tag{E.35}$$

For the torsional actuator the value of the maximum angle is

$$\theta_{\max} = \sin^{-1}\left(\frac{d}{L}\right) = 23.58°. \tag{E.36}$$

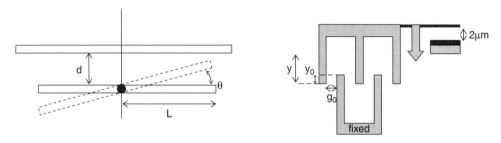

Figure E.7   Problem 5.11

The required voltage to close the torsional plate to touch ground occurs at the pull-in at $\theta = 0.44\theta_{max}$. Then the torsional moment equals the opposing moment at the hinge: $M = k_\theta\theta$. From equation (5.72),

$$M = \frac{\epsilon W V^2}{2\sin\theta\tan\theta}\left(\ln\left(1 + \frac{L\tan\theta}{d}\right) + \frac{d}{d + L\tan\theta} - 1\right) = k_\theta\theta. \tag{E.37}$$

The result is $V = 12.3$ V. The energy consumed from the source is

$$E = V^2 C_{max}, \tag{E.38}$$

where C_{max} is the maximum capacity of the torsional actuator. Taking into account equation (5.67) and with a change of variable $z = d + x\tan\theta$, with $z_1 = d + x_1\tan\theta$ and $z_2 = d + x_2\tan\theta$,

$$dC = \frac{\epsilon_0 W}{z\tan\theta}dz. \tag{E.39}$$

Integrating yields

$$C = \int_{x_1}^{x_2} dC = \frac{\epsilon_0 W}{\tan\theta}\int_{z_1}^{z_2}\frac{dz}{z} = \frac{\epsilon_0 W}{\tan\theta}\ln\left(\frac{d + x_2\tan\theta}{d + x_1\tan\theta}\right). \tag{E.40}$$

With $b = x_2 - x_1$ and $x_1 \ll x_2$,

$$C = \frac{\epsilon_0 W}{\tan\theta}\ln\left(1 + \frac{b\tan\theta}{d}\right). \tag{E.41}$$

For $\theta = \theta_{max}$, $C = C_{max} = 3.05 \times 10^{-16}$ F and the energy consumed is $E = 5.13 \times 10^{-12}$ J.

Problem 5.12

We have a comb actuator and we want to calculate the number of fingers required to have a displacement of 20 μm when the applied voltage is 20V. The data values are $y_0 = 200$ μm, $t = 2$ μm, $g_0 = 1$ μm and $k = 0.01$ N/m.

Taking into account that

$$y = y_0 + \frac{N\epsilon_0 V^2 t}{kg_0}, \tag{E.42}$$

isolating N yields

$$N = \frac{kg_0(y - y_0)}{\epsilon_0 V^2 t} = 254. \tag{E.43}$$

Problem 5.13

We have an accelerometer based on the measurement of displacement described in Section 5.12. If we set a target value of $\omega_0 = 2\pi \times 10^3$ rad/s, simulate and plot the transfer function $|H(j\omega)|$ as a function of the frequency.

The transfer function is

$$H(s) = \frac{X(s)}{A_{ext}(s)} = \frac{m}{ms^2 + bs + k} = \frac{1}{s^2 + \frac{b}{m}s + \frac{k}{m}} = \frac{1}{s^2 + 2\zeta\omega_0 s + \omega_0^2}, \qquad (E.44)$$

and

$$|H(j\omega)| = \left| \frac{X(j\omega)}{A_{ext}(j\omega)} \right| = \frac{1}{\sqrt{(\omega_0^2 - \omega^2)^2 + (2\zeta\omega_0\omega)^2}}. \qquad (E.45)$$

We write a PSpice file as follows:

```
*********problem513
vin 1 0 ac 1
.param pi=3.1416
.param zeta=0.1
.param omega0=6.28e3
eaccelerometer 10 0 laplace v(1) =
1/(s^2+2*zeta*omega0*s+omega0^2)
.ac dec 1000 1e0 1e5
.probe
.end
```

and the results are shown in Figure E.8.

As can be seen, the value of the transfer function when $\omega \to 0$ is

$$|H(j\omega)|_{\omega \to 0} = \frac{1}{\omega_0^2} = 25.3 \times 10^{-9} \, \text{m/ms}^{-2}. \qquad (E.46)$$

Problem 5.14

With the same data as in Problem 5.13, find the transient response of the accelerometer to a pulse of 10 ms^{-2} and plot it as a function of time.

We can use the same description of the transfer function by means of the Laplace source in the PSpice code of Problem 5.13, but running a transient simulation instead of an AC simulation.

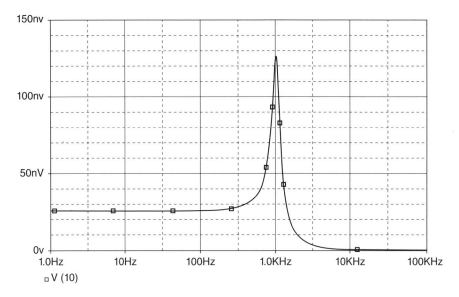

Figure E.8   Results Problem 5.13

We need to define a pulse source for Vin instead of the AC source that we used in Problem 5.13. This is seen in the following code:

```
*********problem514
Vin 1 0 PULSE(0V 10V 0s 10us 10us 50ms 100ms)
.param pi=3.1416
.param zeta=0.1
.param omega0=6.28e3
eaccelerometer 10 0 laplace v(1) = 1/(s^2+2*zeta*omega0*s+omega0^2)
.TRAN 1us 100ms 0s 1us
.PROBE
.END
```

Here we have used a 50 ms wide pulse with 10 μs rise and fall times to avoid convergence problems and zero delay. The result is shown in Figure E.9. We can see that the response of the accelerometer is underdamped and there is quite a high overshoot. Full settling or the response is reached at approximately 8 ms.

Problem 5.15

For the same value of $\omega_0 = 2\pi \times 10^3$ rad/s as in Problem 5.13, find the critical damping condition and plot the transient response to an acceleration pulse of 10 ms^{-2} as a function of time.

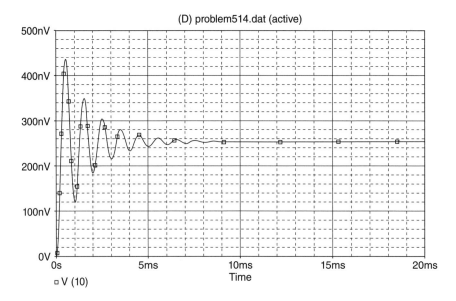

Figure E.9   Results of Problem 5.14

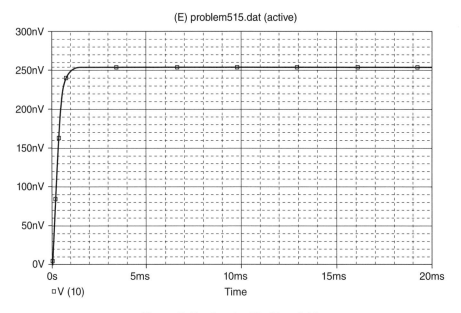

Figure E.10   Result of Problem 5.15

The critical damping condition is reached when the poles of the transfer function have zero imaginary part. The roots of the denominator of the transfer function are found to be

$$s = -\zeta\omega_0 \pm \omega_0\sqrt{\zeta^2 - 1}. \tag{E.47}$$

The critical damping condition is thus reached for $\zeta = 1$. Simulating the same file as in Problem 5.14 but with this value for ζ, the result shown in Figure E.10. It can be seen that the settling time of the steady response is reduced to approximately 1 ms.

F

Chapter 6 Solutions

Problem 6.1

Show that the factor Q is equal to the resonant frequency divided by the bandpass.

The amplitude of the deflection is related to the frequency by

$$\hat{x} = \frac{F_0/m}{\sqrt{\left(\omega_0^2 - \omega^2\right)^2 + (2\zeta\omega_0\omega)^2}}, \tag{F.1}$$

and the amplitude has a maximum given by

$$\hat{x}(\omega = \omega_r) = \frac{F_0}{2k\zeta\sqrt{1 - \zeta^2}}. \tag{F.2}$$

Normalizing to the low-frequency value,

$$\hat{x}(\omega = \omega_r)_{\text{norm}} = \frac{F_0}{\frac{F_0}{m}2k\zeta\sqrt{1 - \zeta^2}} = \frac{1}{2\omega_0\zeta(1 - \zeta^2)}, \tag{F.3}$$

and the amplitude at a frequency also normalized to the the value F_0/m is given by

$$\hat{x} = \frac{1}{\sqrt{\left(\omega_0^2 - \omega^2\right)^2 + (2\zeta\omega_0\omega)^2}}. \tag{F.4}$$

We want the frequencies where the normalized amplitude equals the normalized maximum amplitude divided by $\sqrt{2}$, hence,

$$\frac{1}{2\omega_0\zeta(1 - \zeta^2)}\frac{1}{\sqrt{2}} = \frac{1}{\sqrt{\left(\omega_0^2 - \omega^2\right)^2 + (2\zeta\omega_0\omega)^2}} \tag{F.5}$$

Understanding MEMS: Principles and Applications, First Edition. Luis Castañer.
© 2016 John Wiley & Sons, Ltd. Published 2016 by John Wiley & Sons, Ltd.
Companion Website: www.wiley.com/go/castaner/understandingmems

or

$$(2\omega_0\zeta(1-\zeta^2)\sqrt{2})^2 = \left(\omega_0^2-\omega^2\right)^2 + (2\zeta\omega_0\omega)^2, \tag{F.6}$$

and we end up with a quadratic equation

$$\omega^4 + 2\omega_0^2\omega^2(2\zeta^2-1) + \omega_0^4(1-8\zeta^2(1-\zeta^2)^2) = 0. \tag{F.7}$$

As this is a second-order equation in $y = \omega^2$, the two roots are

$$y_1 = (1-2\zeta^2) + 2\zeta\sqrt{1-\zeta^2} \tag{F.8}$$

and

$$y_2 = (1-2\zeta^2) - 2\zeta\sqrt{1-\zeta^2}. \tag{F.9}$$

Taking the first two terms of the Taylor expansion of $\sqrt{1-\zeta^2}$,

$$\sqrt{1-\zeta^2} \simeq 1 - \frac{\zeta^2}{2}, \tag{F.10}$$

we obtain

$$y_1 \simeq (1-2\zeta^2) + 2\zeta\left(1-\frac{\zeta^2}{2}\right) \simeq 1 + 2\zeta - 2\zeta^2 - \zeta^3 \simeq 1 + 2\zeta \tag{F.11}$$

and

$$y_2 \simeq (1-2\zeta^2) - 2\zeta\left(1-\frac{\zeta^2}{2}\right) \simeq 1 - 2\zeta - 2\zeta^2 - \zeta^3 \simeq 1 - 2\zeta. \tag{F.12}$$

Then the two cut-off angular frequencies are

$$\omega_1^2 \simeq \omega_0^2(1+2\zeta), \tag{F.13}$$
$$\omega_2^2 \simeq \omega_0^2(1-2\zeta). \tag{F.14}$$
$$\tag{F.15}$$

We can write

$$\omega_1^2 - \omega_2^2 = (\omega_1+\omega_2)(\omega_1-\omega_2) \tag{F.16}$$

and

$$\omega_1^2 - \omega_2^2 = 4\zeta\omega_0, \tag{F.17}$$

as

$$\omega_1 + \omega_2 = \omega_0(\sqrt{1+2\zeta} + \sqrt{1-2\zeta}) \simeq 2\omega_0, \tag{F.18}$$

and we obtain

$$\omega_1 - \omega_2 = \frac{4\zeta\omega_0^2}{2\omega_0} = 2\zeta\omega_0. \tag{F.19}$$

The quality factor Q is given by

$$Q = \frac{\omega_0}{\omega_1 - \omega_2} \simeq \frac{1}{2\zeta}. \tag{F.20}$$

Problem 6.2

With the same data values as in Example 6.3.2, calculate the value of Q from the results of the PSpice simulation.

From the simulations shown in Figure 6.1 and using the cursor utility, we first find the maximum value of the amplitude and its frequency. We consider this frequency as the resonance frequency. We then find the value of the maximum divided by $\sqrt{2}$ and then we find the values of the two cut-off frequencies corresponding to this value of the amplitude: one smaller that the resonance and the other larger than the resonance. We next find the quotient of the resonance frequency divided by the difference of the two cut-off frequencies. We do these calculations for all three values of the damping coefficient ζ. The results are summarized in Table F.1.

Table F.1   Summary of results for Problem 6.2

ζ ($\times 10^{-4}$)	f_{max} (MHz)	X_{max} ($\times 10^{-12}$)	$X_{max}/\sqrt{2}$ ($\times 10^{-12}$)	f_1 (MHz)	f_2 (MHz)	Δf (kHz)	$Q = \frac{f_{max}}{\Delta f}$
1	999.38	126.96	89.775	999.485	999.285	200.8	4976.9
2	999.38	63.427	44.84	999.587	999.183	404	2473
5	999.38	25.399	17.95	999.8831	998.886	9971	1000.2

Problem 6.3

We have a vibrating mass of a plate $500 \times 500~\mu m^2$ made of silicon with a density of $2339~kg/m^3$, and the plate thickness is $h = 0.2~\mu m$. The initial gap is $g_0 = 1~\mu m$. We assume that the applied bias voltage makes the plate deflect by $x_B = \frac{1}{4}g_0$, corresponding to a bias voltage of 10 V. We assume a damping coefficient of $b = 3 \times 10^{-3}$. Calculate the values of the components of the equivalent LCR circuit.

From equations (6.53), (6.54) and (6.55),

$$L = m\frac{(g_0 - x_B)^4}{(\epsilon_0 A V_B)^2} = 7.43 \times 10^{-4}. \tag{F.21}$$

The equivalent stiffness constant can be calculated from the equilibrium conditions:

$$\frac{\epsilon_0 A V_B^2}{2(g_0 - x_B)^2} - k x_B = 0, \tag{F.22}$$

from which $k = 7.86 \times 10^3$. The capacitance C is given by

$$C = \left(k\frac{(g_0 - x_B)^4}{(\epsilon_0 A V_B)^2} - \frac{g_0 - x_B}{\epsilon_0 A} \right)^{-1} = 1.96 \times 10^{-14}\ \text{F}, \tag{F.23}$$

and the resistance R_M by

$$R_M = b\frac{(g_0 - x_B)^4}{(\epsilon_0 A V_B)^2} = 19.3\ \text{k}\Omega. \tag{F.24}$$

The gap capacitance is

$$C_g = \frac{\epsilon_0 A}{g_0 - x_B} = 3 \times 10^{-15}. \tag{F.25}$$

Problem 6.4

A laterally resonating gravimetric sensor is described in [40]. The proof mass is made out of the patterning of the 5 μm device layer of a SOI wafer. The other two dimensions of the mass are 160 μm and 30 μm. The density of silicon is 2339 kg/m³. From the reference, the resonant frequency can be identified as $\omega_0 = 245.5$ kHz. (1) Calculate the value of the effective stiffness constant of the flexures supporting the proof mass. (2) Calculate the value of frequency shift that can be expected if a polystyrene bead of 550 pg is deposited on top of the proof mass.

---

(1) We have

$$\omega_0 = \sqrt{\frac{k_{\text{eff}}}{m}}, \tag{F.26}$$

that is,

$$k_{\text{eff}} = \omega_0^2 m = 5.43 \times 10^{-4}\ \text{N/m}. \tag{F.27}$$

(2) Taking into account Example 6.6.2, the change in frequency is

$$\frac{\Delta\omega}{\omega} = -\frac{1}{2}\frac{\Delta m}{m}. \tag{F.28}$$

The proof mass is

$$m = 160 \times 10^{-6} \times 30 \times 10^{-6} \times 5 \times 10^{-6} \times 2339 \, \text{kg} = 5.613 \times 10^{-11} \, \text{kg}. \quad \text{(F.29)}$$

Applying equation (F.28),

$$\Delta\omega = -\frac{1}{2} \frac{550 \times 10^{-15}}{5.613 \times 10^{-11}} \, 45.5 \times 10^3 = -1.202 \times 10^3 \, \text{Hz}. \quad \text{(F.30)}$$

The minus sign indicates that the increasing mass produces a negative shift of the resonance frequency

---

Problem 6.5

Calculate the value of the free vibration frequency, the value of the damped vibration frequency and the frequency at which the maximum amplitude of deflection occurs for a cantilever with the following data: $W = 10 \times 10^{-6}$ *m,* $h = 1.5 \times 10^{-6}$ *m,* $L = 200 \times 10^{-6}$ *m,* $g = 9.8$ *m/s^2,* $\rho = 2329$ *kg/m^3,* $E = 130$ *GPa,* $\zeta = 0.1$.

We assume that the cantilever is subject to its own weight. The mass of the cantilever is

$$m = WLhg = 6.9 \times 10^{-12} \, \text{kg}, \quad \text{(F.31)}$$

and the stiffness constant is equal to the weight divided by the deflection at the tip. For a cantilever subject to its own weight, the deflection is

$$v = \frac{\rho_m g W h}{24 EI} x^2 (x^2 + 6L^2 - 4Lx), \quad \text{(F.32)}$$

and at $x = L$,

$$v(L) = \frac{3 \rho_m g W h L^4}{24 EI}. \quad \text{(F.33)}$$

Then the value of the stiffness (for the rectangular cross-section cantilever) is

$$k = \frac{mg}{v(L)} = \frac{\rho_m g W h L}{v(L)} = \frac{2}{3} \frac{E W h^3}{L^3} = 0.12 \, \text{N/m}. \quad \text{(F.34)}$$

(1) The free vibration frequency is

$$\omega_0 = \sqrt{\frac{k}{m}} = 1.3187 \times 10^6 \, \text{rad/s}. \quad \text{(F.35)}$$

(2) The damping frequency is

$$\omega_d = \omega_0 \sqrt{1 - \zeta^2} = 0.994\omega_0 = 1.3180 \times 10^6 \ \text{rad/s}. \tag{F.36}$$

(3) The frequency ω_{max} is given by

$$\omega_r = \omega_0 \sqrt{1 - 2\zeta^2} = 0.989\omega_0 = 1.303 \times 10^6 \ \text{rad/s}. \tag{F.37}$$

Problem 6.6

In a tuning fork gyroscope similar to the one described in Section 6.7, the comb drive is composed of 250 fingers and the main parameter values are: $t_{comb} = 2 \ \mu m$, $g_{0D} = 1 \ \mu m$, $x_0 = 100 \ \mu m$, $\omega_x = 2\pi \times 15 \times 10^3 \ rad/s$, $V_{CC} = 30 \ V$, $Q_x = Q_y = 10\,000$. Calculate the sensitivity.

The AC component of the force in the drive electrode is

$$F_{xAC} = \frac{2N\epsilon_0 t_{comb}}{g_{0D}} V_{CC} v_d = \frac{2 \times 250 \times 8.85 \times 10^{-12} \times 2 \times 10^{-6}}{1 \times 10^{-6}} 30 v_d$$
$$= 2.65 \times 10^{-7} v_d. \tag{F.38}$$

For an AC driving voltage of $vd = 0.5$ V, the amplitude of the displacement in the sense direction is

$$\hat{y} = \frac{4 \times 250 \times 8.85 \times 10^{-12} \times 2 \times 10^{-6} \times 1 \times 10^4 \times 10^4}{1 \times 10^{-6} \times 234\omega_x} 30 v_d \Omega$$
$$= 1.2 \times 10^{-6} \Omega \tag{F.39}$$

Then the sensitivity 1.2 μm/(rad/s).

Problem 6.7

Calculate the effective stiffness of a resonant cantilever due to the steady-state electrostatic attraction as a function of the bias position of the cantilever x_B. The dimensions of the cantilever are $500 \times 10 \times 2 \ \mu m$, the initial gap is $g_0 = 5 \ \mu m$ and the elastic stiffness $k = 1 \ N/m$.

The effective stiffness is given by

$$k_{eff} = k \left(1 - \frac{2x_B}{g_0 - x_B} \right) \tag{F.40}$$

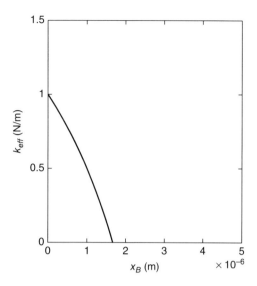

Figure F.1 Results for Problem 6.7

which, for the parameter values of the problem, gives

$$k_{eff} = 1 - \frac{2x_B}{5 \times 10^{-6} - x_B}. \tag{F.41}$$

This result is plotted in Figure F.1. As can be seen, the stiffness is reduced as the value of x_B increases, and when $x_B = \frac{1}{3}g_0$ the value reaches zero. At this point the pull-in instability is also reached and thus the resonance frequency has no physical meaning.

Problem 6.8

A double-clamped bridge has a first resonance given by [19, p. 218]

$$\omega(0) = 22.27\sqrt{\frac{EI}{\rho_m WtL^4}}. \tag{F.42}$$

When the bridge is subject to an axial force, the resonance frequency shifts due to the axial force F, as follows [41]:

$$\omega(F) = \omega(0)\sqrt{1 + \gamma\frac{FL^2}{12EI}}, \tag{F.43}$$

where $\gamma = 0.2949$. If the beam is subject to a pressure $P = 10$ kPa, calculate the shift in resonance frequency. The bridge has a length of $L = 200$ μm, a width $W = 20$ μm and thickness

$t = 2\ \mu m$. ρ_m is the density of the bridge material, assumed to be silicon: $\rho_m = 2340\ kg/m^3$. Young's modulus is $E = 169\ GPa$.

We calculate first the moment of inertia,

$$I = \frac{Wt^3}{12} = \frac{20 \times 10^{-6} \times (2 \times 10^{-6})^3}{12} = 1.33 \times 10^{-23}, \tag{F.44}$$

and then the value of $\omega(0)$ from equation (F.42),

$$\omega(0) = 2.78 \times 10^6\ rad/s. \tag{F.45}$$

To find the value of the axial force F, we use equation (3.88), where the axial stress in a doubly supported beam was calculated at the edges of the beam and as a function of the position in the bridge. We will take the value of the axial stress at $x = L$, and then

$$F = \sigma_x Wt. \tag{F.46}$$

From equation (3.88) at $x = L$,

$$\sigma_x = \frac{PWtL^2}{24I}. \tag{F.47}$$

The resonance frequency is

$$\omega(F) = \omega(0)\sqrt{1 + 0.2949\frac{2PL^4}{t^4E}} = 1.0172\omega(0). \tag{F.48}$$

The frequency shift is

$$\Delta\omega = \omega(F) - \omega(0) = 0.0172\omega(0) = 46.44\ krad/s. \tag{F.49}$$

Problem 6.9

A resonant accelerometer is based on a proof mass attached to two double-ended tuning fork resonators [42] Find the expected frequency change for an acceleration of 10 g. The dimensions of the beam are $L = 450\ \mu m$, $W = 6\ \mu m$ and $t = 10\ \mu m$. Young's modulus is $E = 169\ GPa$. The resonant frequency in the absence of acceleration is $f(0) = 81.978\ kHz$.

We use equation (F.48), and we assume that the axial force is $F = ma$, where a is the acceleration. Taking into account the parameter values of the problem, we have

$$\omega(F) = \omega(0)\sqrt{1 + 0.2949\frac{2PL^4}{t^4 E}} = 1.44 \times 10^{-5}\omega(0). \tag{F.50}$$

For $\omega(0) = 2\pi f(0)$, the angular frequency shift is

$$\Delta\omega = \omega(F) - \omega(0) = 7.46\,\text{rad/s}. \tag{F.51}$$

G

Chapter 7 Solutions

Problem 7.1

A drop of water has a contact angle on top of a surface of $\theta_0 = 110°$. We know that $\gamma_{SV} = 10.1 \times 10^{-3}$ N/m and $\gamma_{LV} = 72.8 \times 10^{-3}$ N/m. The bottom conductive substrate is covered by a 80 nm thick layer of Si_3N_4 having a relative permittivity of $\epsilon_r = 7.5$. Calculate the value of γ_{SL}. Calculate the radius of the base circle of a droplet of 20 μl, assuming that the shape can be approximated by a spherical cap. Calculate the voltage that must be applied for the contact angle to have a value of 75°. Calculate the new value of the radius of the base circle.

The initial contact angle is

$$\cos\theta_0 = \frac{\gamma_{SV} - \gamma_{SL}}{\gamma_{LV}} = \frac{19.1 \times 10^{-3} - \gamma_{SL}}{72.8 \times 10^{-3}} = \cos 100°, \tag{G.1}$$

yielding

$$\gamma_{SL} = 0.0439. \tag{G.2}$$

The radius of the base circle is

$$r = \left(\frac{3vol}{\pi}\right)^{1/3} \frac{\sin\theta}{(2 - 3\cos\theta + \cos^3\theta)^{1/3}} \tag{G.3}$$

$$= \left(\frac{3 \times 20 \times 10^{-9}}{\pi}\right)^{1/3} \frac{\sin 110°}{(2 - 3\cos 110° + \cos^3 110°)^{1/3}} = 1.73 \text{ mm}. \tag{G.4}$$

When the voltage is applied the new value of the contact angle is

$$\cos\theta = \cos\theta_0 + \frac{C_A V^2}{2\gamma_{LV}} = 0 \tag{G.5}$$

$$\cos 75° = \cos 110° + \frac{7.5 \times 8.85 \times 10^{-12} V^2}{2 \times 72.8 \times 10^{-3} \times 80 \times 10^{-9}}, \tag{G.6}$$

Understanding MEMS: Principles and Applications, First Edition. Luis Castañer.
© 2016 John Wiley & Sons, Ltd. Published 2016 by John Wiley & Sons, Ltd.
Companion Website: www.wiley.com/go/castaner/understandingmems

yielding $V = 10.6$ V. The new value of the radius of the base circle is

$$r = \left(\frac{3 \times 20 \times 10^{-9}}{\pi}\right)^{1/3} \frac{\sin 75°}{(2 - 3\cos 75° + \cos^3 75°)^{1/3}} = 1.82 \, \text{mm}. \qquad \text{(G.7)}$$

Problem 7.2

A drop of water on top of a substrate forms a contact angle of 110° and when a 10 V voltage is applied, the contact angle reduces to 90°. Calculate the surface tension γ_{LV} for a capacitance per unit area of $C_A = 1 \times 10^{-4}$ F/m².

The surface tension is

$$\gamma_{LV} = \frac{C_A V^2}{2(\cos\theta - \cos\theta_0)} = 0.0146. \qquad \text{(G.8)}$$

Problem 7.3

One way to avoid direct contact on the liquid using a needle or an upper electrode is to make use of a capacitive coupling between electrodes in the bottom substrate as shown in Figure 7.6. The dielectric has a thickness $d = 200$ nm, $\gamma_{LV} = 72.8 \times 10^{-3}$, the initial contact angle is $\theta_0 = 110°$, and $\epsilon_r = 2.1$. Write the relationship between the voltages V_S and V_L. Calculate the value of the contact angle after a voltage $V_S = 10$ V is applied.

The two capacitors form a voltage divider and, as they are connected in series, the charge in both has to be the same,

$$(V_S - V_L)C_S = V_L C_G. \qquad \text{(G.9)}$$

Then

$$V_L = V_S \frac{C_S}{C_G + C_S}, \qquad \text{(G.10)}$$

taking into account that

$$C_S = \frac{\epsilon_0 \epsilon_r A_S}{d} \quad \text{and} \quad C_G = \frac{\epsilon_0 \epsilon_r A_G}{d}, \qquad \text{(G.11)}$$

where A_S and A_G are the area of the electrodes of the source and ground, respectively. Thus

$$V_L = V_S \frac{A_S}{A_G + A_S}. \qquad \text{(G.12)}$$

If the electrodes areas are $A_G = 1.2 \times 10^{-7}$ m^2 and $A_S = 2.25 \times 10^{-8}$ m^2, the voltage at the liquid is $V_L = 1.58$ V. At the periphery of the drop the voltage across the dielectric is $V_S - V_L = 10$ V $- 1.58$ V $= 8.42$ V. Finally, the contact angle is given by

$$\cos \theta = \cos \theta_0 + \frac{C_A(V_S - V_L)^2}{2\gamma_{LV}} = \cos 110° + \frac{\frac{2.1 \times 8.85 \times 10^{-12}}{200 \times 10^{-9}}(1.58)^2}{2 \times 72.8 \times 10^{-3}}, \qquad (G.13)$$

leading to $\theta = 107.9°$.

Problem 7.4

With the same data as in Problem 7.3, we would like to change the design of the device in order to make the change in contact angle larger. For this reason we reduce the thickness of the dielectric to $d = 50$ nm. Calculate the ratio of the two areas A_S/A_G to achieve a change of $10°$ in the contact angle after $V_S = 10$ V is applied.

We have

$$V_S - V_L = \sqrt{\frac{2\gamma_{LV}(\cos \theta - \cos \theta_0)}{C_A}} = 8.11 \text{ V}. \qquad (G.14)$$

V_S and V_L are also related by

$$V_S - V_L = V_S \left(1 - \frac{A_S}{A_G + A_S}\right). \qquad (G.15)$$

Then $A_S/A_G = 0.23$.

Problem 7.5

We would like to compare the electro-osmotic flow with pressure-driven flow. For the same flow Q in both cases, calculate the correspondence between the electric field value and the pressure gradient value for the following data: $V_0 = 100$ mV, $W = 100$ μm, $h = 20$ μm, $\epsilon_0 = 8.85 \times 10^{-12}$ F/m and $\epsilon_r = 80$.

The electro-osmotic flow is given by

$$F = \frac{1}{\mu} Wh\epsilon_0\epsilon_r V_0 \mathbf{E}, \qquad (G.16)$$

where μ is the viscosity, V_0 is the surface potential and $\mathbf{E}$ the longitudinal electric field. We also know that the pressure-driven flow is given by

$$F = -\frac{1}{12\mu} Wh^3 \frac{\Delta P}{\Delta x}. \tag{G.17}$$

To achieve the same flow, the correspondence between the electric field required in an electro-osmotic flow and the pressure gradient in a pressure-driven flow is found by making the two equations above equal:

$$\frac{\Delta P}{\Delta x} = \frac{12\epsilon V_0}{h^2}\mathbf{E}. \tag{G.18}$$

With the data for the problem,

$$\frac{\Delta P}{\Delta x} = 212.4\mathbf{E}. \tag{G.19}$$

Problem 7.6

For the same data as in Problem 7.5, calculate the pressure gradient required to sustain a flow of 200 μl/min in a PDMS channel 1000 μm long. The liquid viscosity is 0.000749 Pa s.

We have

$$F = -\frac{1}{12\mu} Wh^3 \frac{\Delta P}{\Delta x} = 9.38 \times 10^{-16} \frac{\Delta P}{\Delta x}. \tag{G.20}$$

The flow of 200 μl/min $= 3.33 \times 10^{-9}\,\mathrm{m^3/s}$, so

$$\frac{\Delta P}{\Delta x} = 3.5 \times 10^6\,\mathrm{Pa/m}. \tag{G.21}$$

Problem 7.7

We have a 100 × 100 μm cross-section channel of length L = 1 mm. We want to deliver a flow of 100 μl/hour. Calculate the pressure difference that must be applied. The liquid viscosity is μ = 8.9 × 10⁻⁴ Pa s.

The pressure difference is given by

$$\Delta P = -\frac{12\mu L}{Wh^3}F = \frac{12 \times 8.9 \times 10^{-4} \times 10^{-3}}{100 \times 10^{-6} \times 100^3 \times 10^{-18}}F, \tag{G.22}$$

and the value of F is

$$F = 100 \frac{\mu l}{s} \times \frac{11}{1 \times 10^6 \, \mu l} \frac{1 \, m^3}{10001} = 1 \times 10^{-7} \, m^3/s, \tag{G.23}$$

therefore $\Delta P = 1.068$ kPa.

Problem 7.8

Write Matlab code to solve the electrowetting equations (7.86) and (7.93)

The problem has three state variables: the contact angle, θ; the first derivative of the contact angle, $d\theta/dt$; and the charge per unit area, q_A. The three are the elements of a vector y given by

$$\begin{pmatrix} y(1) \\ y(2) \\ y(3) \end{pmatrix} = \begin{pmatrix} \theta \\ \frac{d\theta}{dt} \\ q_A \end{pmatrix}. \tag{G.24}$$

and the vector of the derivatives of the vector y, called $dydx$, is given by

$$dydx = \begin{pmatrix} dydx(1) \\ dydx(2) \\ dydx(3) \end{pmatrix} = \begin{pmatrix} \frac{d\theta}{dt} \\ \frac{d^2\theta}{dt^2} \\ \frac{dq_A}{dt} \end{pmatrix} \tag{G.25}$$

$$= \begin{pmatrix} y(2) \\ -\frac{1}{\Omega}\frac{d\Omega}{d\theta}y(2)^2 + y(2)\left(\frac{-\gamma_{LV}\sin y(1)}{\zeta a\Omega} - \frac{2y(3)^2}{C_A \zeta r} \right) + \frac{q_A}{C_A \zeta a\Omega RA}\left(\frac{y(3)}{C_A} - V_s \right) \\ \frac{V_s}{RA} - y(3)\left(-\frac{2a\Omega}{r}y(2) + \frac{1}{ARC_A} \right) \end{pmatrix}.$$

The functions involved in equation (G.25) are defined in Chapter 7, Ω and α in equation (7.85), r in equation (7.67) and A in equation (7.88); C_A is the capacitance per unit area and R is the source electrical resistance.

```
tmax=10;
tspan=linspace(0,tmax,n);
[t,y]=ode15s(@contact,tspan,[theta0,0,0]);
function dy=contact(t,y)
dy=zeros(3,1);
dy(1)=y(2);
dy(2)=-DLOmega(y(1))*(y(2))^2+y(2)*(-gammaLV*sin(y(1))/
(zeta*alpha*Omega(y(1)))-2*(y(3))^2/(ca*zeta*radio(y(1))))+
```

```
y(3)/(ca*zeta*alpha*R*Omega(y(1))*area(y(1)))*(y(3)/ca-Vs);
dy(3)=Vs/(R*area(y(1)))-y(3)*(-2*alpha*Omega(y(1))/
radio(y(1))*y(2)+(1/(area(y(1))*R*ca)));
```

with

```
function DLO = DLOmega(q)
DLO=-(2.*sin(q).*cos(q).*(1-cos(q)))./
(2-3.*cos(q)+(cos(q)).^3);
```

and

```
function a = area(n)
a=pi.*(alpha.*sin(n)./(2-3.*cos(n)+(cos(n)).^3).^(1/3)).^2;
```

and (for the radius of the bottom area of the drop)

```
function r = radio(w)
r=alpha.*sin(w)./(2-3.*cos(w)+(cos(w)).^3).^(1/3);
```

and (for the function Ω)

```
function O = Omega(m)
O=(1-cos(m)).^2./(2-3.*cos(m)+(cos(m)).^3).^(4/3);
```

Problem 7.9

Calculate the contact angle from maximum to minimum after applying a 70 V voltage using the same data as in Example 7.6.1, but for several values of the friction coefficient $\zeta = 1, 20, 50$. Calculate approximately from the plots the time required for the change.

Using the Matlab code in Problem 7.8, we find the results in Table G.1. As can be seen, the change in the contact angle is independent of the value of the friction coefficient, whereas the time required is not.

Table G.1   Results for Problem 7.8

ζ	$\Delta\theta$ (rad)	Δt (s)
1	1.07	$\sim 6 \times 10^{-2}$
10	1.07	$\sim 6 \times 10^{-1}$
20	1.07	~ 1
50	1.07	~ 5

Problem 7.10

Calculate the change in contact angle for the same data as in Example 7.7.1, but for several values of the applied voltage, $V = 20, 40, 60, 75$ V.

The results are given in Table G.2. As can be seen, the change in contact angle increases rapidly with an increase in the value of the applied voltage, whereas the time to change drops slightly at the beginning and then rises again.

Table G.2   Results for Problem 7.10

V	$\Delta\theta$ (rad)	Δt (s)
20	0.08	~1
40	0.32	~0.7
60	0.735	~0.65
75	1.363	~4

Problem 7.11

Dielectrophoresis is used to levitate, classify and sort nanoparticles. In [50] a series of interdigitated electrodes are used to make latex particles levitate. Calculate the required value of the Clausius–Mossotti factor for a particle of 3 μm radius suspended in water to counterbalance the gravitational force, if the estimated value of the electric field gradient squared is 10^{12} V²/m³.

The gravitational force on the bead suspended in water is given by

$$F = \frac{4}{3}\pi r^3(\rho_{mm} - \rho_{mp})g = 5.5 \times 10^{-14} \text{ N}. \tag{G.26}$$

where ρ_{mm} is the medium density and ρ_{mp} is the particle density. As the dielectrophoretic force is

$$F_{\text{DEP}} = 2\pi r^3 \epsilon_m K \nabla |\mathbf{E}|^2 = 1.2 \times 10^{-13} K, \tag{G.27}$$

the required value of the Clausius–Mossoti factor is $K = 0.458$, within the range of theoretical values, which is between 1 and −0.5, depending on the frequency.

H

Chapter 8 Solutions

Problem 8.1

We have a die of single-crystal silicon with a platinum resistor patterned on its surface, as shown in Figure H.1. The resistor has a value of $R(350\,K) = 100\,\Omega$, thickness $t_R = 50nm$ and width of $W_R = 5\,\mu m$. We also know that the resistivity of the platinum film at 350 K is $\rho = 1.281 \times 10^{-7}\,\Omega\,m$. Assuming that the heat is generated from the dissipation of power when the resistor is fed with a constant current I, and that it is evenly distributed across the silicon die, write the equation of the resistance of the Δx shown in Figure H.1.

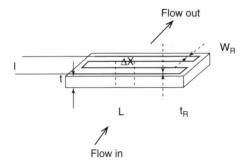

Figure H.1 Problem 8.1

We know that the resistance is defined as

$$R = \rho \frac{L_R}{t_R W_R}. \tag{H.1}$$

If the power dissipation is distributed evenly and as the current is constant, we have to conclude that this is equivalent to distributing the resistor evenly, so for a length Δx,

$$dR = \rho \frac{\Delta x}{t_R W_R} = 521.4 \times 10^3 \Delta x. \tag{H.2}$$

Understanding MEMS: Principles and Applications, First Edition. Luis Castañer.
© 2016 John Wiley & Sons, Ltd. Published 2016 by John Wiley & Sons, Ltd.
Companion Website: www.wiley.com/go/castaner/understandingmems

In order to enter this parameter into the heat equation we can write it as

$$dR = R\frac{\Delta x}{L}. \tag{H.3}$$

In this way, although the resistance need not necessarily cover all the area nor the length of the heated volume, it can be assumed to be evenly distributed.

Problem 8.2

Write the heat equation for the geometry shown in Figure H.1, with dR as found in Problem 8.1. Assume that the resistor takes a value of $R = 100\,\Omega$ at a temperature $T_{HOT} = 350\,K$.

The main parameters are

$$A_{cond} = Wt \tag{H.4}$$

and

$$A_{conv} = 2\Delta x(W + t), \tag{H.5}$$

and dR can be written as

$$dR = 100\frac{\Delta x}{L}. \tag{H.6}$$

Thus

$$\dot{Q}_{gen} = I^2 dR = I^2 \times 100 \times \frac{\Delta x}{L}. \tag{H.7}$$

Substituting in the heat equation,

$$k\frac{d^2T}{dx^2}WT\Delta x = -100I^2\frac{\Delta x}{L} + 2h\Delta x(W + t)\Delta T \tag{H.8}$$

or

$$\frac{d^2T}{dx^2} = -I^2\frac{100}{kWtL} + \frac{2h\Delta x(W + t)}{kWt}\Delta T. \tag{H.9}$$

Problem 8.3

For the heat equation found in Problem 8.2, find suitable boundary conditions if the two ends of the silicon die are attached by means of two supports of the same cross-section but with different thermal conductivity, as shown in Figure H.2.

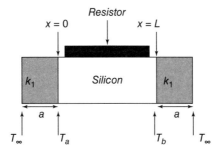

Figure H.2 Problem 8.3

Assuming that the two low thermal conductivity supports have a constant heat flux q_1 (and thus neglecting any convection losses in them), we can write

$$q_1 = -k_1 \frac{dT}{dx} = -k_1 \frac{T_a - T_\infty}{a},$$ (H.10)

where a is the length of the low conductivity support. The boundary condition in this case, as the temperature of the interface is not fixed, is a boundary condition of the second kind (also known as a Neumann condition), which establishes that we have the same heat rate on the two sides of the interface. As the cross-section is the same,

$$q_s|_{x=0} = -k_s \frac{dT}{dx}\bigg|_{x=0} = -k_1 \frac{T_a - T_0}{a},$$ (H.11)

where k_s is the thermal conductivity of the silicon die, T_a is the temperature at $x = 0$, and T_0 is the temperature at the other end of the support. The boundary condition at $x = 0$ can be written as

$$\frac{dT}{dx}\bigg|_{x=0} = \frac{k_1}{k_s} \frac{T_a - T_0}{a}$$ (H.12)

or

$$T_a = T_0 + \frac{k_s}{k_1} a \frac{dT}{dx}\bigg|_{x=0}.$$ (H.13)

Similarly, the boundary condition at $x = L$ can be written as

$$T_b = T_0 - \frac{k_s}{k_1} a \frac{dT}{dx}\bigg|_{x=L}$$ (H.14)

Due to symmetry, $T_a = T_b$.

Problem 8.4

For the same geometry as in Problem 8.3, write Matlab code to solve the differential equation with the boundary conditions of Problem 8.3, assuming that the silicon die has length $L = 1000\,\mu$, thickness $t = 100\,\mu m$ and width $W = 1000\,\mu m$, and for heat convection coefficient $h = 100, 200, 500\ W/m^2C$.

The differential equation found in Problem 8.3 can now be solved with the help of the two boundary conditions we have found. We write the following Matlab code:

```
solinit= bvpinit(linspace(0,L,2000),[To 0])
sol=bvp4c(@problem8_1functionbook,@problem8_1bcbook,solinit);
x=linspace(0,L);
y=deval(sol,x);
mean_temp= trapz(x,y(1,:))/L
```

The heat equation is described by the function

```
function dydx=problem8_1functionbook(x,y)
dydx=[y(2)
    2*h*(t+w)/(k*w*t)*(y(1)-Tair)-I^2*100/(L*k*w*t)];
```

and the boundary conditions

```
function res= problem8_1bcbook(ya,yb)
    res=[ya(1)-(To+k/0.04*50e-6 *ya(2))
    yb(1)-(To-k/0.04*50e-6*yb(2))];
    end
```

where we have assumed that the thermal conductivity of the support/isolation material is 0.04 W/m·K.

We have implemented the process flow shown in Figure H.8 for this example, and the results are shown in Table H.1

In this example we have used equations (H.30), (8.31), (8.32) and (8.33) and assumed that the resistance is $R = 100\,\Omega$ because the control strategy of the loop keeps the temperature of the silicon die constant at 350 K regardless of the value of the convection coefficient. As can be seen in Table H.1, as the speed U increases, greater heating power is required to keep the average temperature at 350 K. Moreover, the power gone into convection $\dot{Q} = hA_{conv}(T_{HOT} - T_\infty)$ also

Table H.1 Summary of the results for Problem 8.4, for $\overline{T} = 350$ K

h	I (mA)	I^2R (mW)	$\dot{Q}_{conv}$ (mW)	U (m/s)
100	13.9	19.3	11	0.82
200	17.3	29.9	22	4.98
500	25.2	63.5	55	53.8

increases. A measurement of the value of the heating power is readily translated into a value for the speed. In MEMS, a typical application is the area of respirators or spirometers, where air flows up to 18 l/s are routinely measured.[1] For a duct of diameter 27 mm this corresponds to an air velocity of $U = 31.4$ m/s which is in the range of the results found in Table H.1.

Problem 8.5

For the same geometry as in Problem 8.4, we would like to assess the effect of the value of the thermal conductivity of the supports, k_s, on the power balance. Check using 0.1 and 1 W/m · K.

We run the same Matlab code as in Problem 8.4 and we find the results shown in Table H.2. These values are plotted in Figure H.3. As can be seen, when the thermal conductance of the supports increases, the total power required increases for the same flow velocity, and the percentage of this power lost to convection is accordingly reduced, so the overall efficiency of the sensor is compromised.

Table H.2 Summary of the results of Problem 8.5, for $\overline{T} = 350K$

k_s	h	I (mA)	I^2R (mW)	$\dot{Q}_{conv}$ (mW)	U (m/s)
0.1	100	17.5	30.6	11	0.82
	200	20.5	42.05	22	4.98
1	100	45.7	208.8	11	0.82
	200	47	220	22	4.98

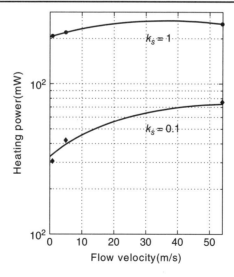

Figure H.3 Results for Problem 8.5

[1] http://www.thormed.com

Problem 8.6

We have a heated volume of silicon with a power of 1 mW using a resistor. Imagine that the silicon volume is assumed to be isothermal. The volume is 1000 μm square by 500 μm thick. This volume is anchored by a thinner and narrower silicon bridge 1000 μm long, 5 μm wide and 2 μm thin. Assume that the convection heat transfer coefficient is h = 100 and that all the surface of the silicon volume is subject to the convective heat transfer. Calculate the silicon volume temperature if the air temperature is $T_\infty = 300$ K, and the frame temperature where the bridge is anchored is 310 K. Calculate the amount of power required to heat the silicon temperature up to 340 K. The thermal conductivity of silicon is assumed to be 133.5 W/m·K.

According to Figure H.4, the thermal circuit equation is

$$\dot{Q} = \dot{Q}_1 + \dot{Q}_{conv} = \frac{T_{si} - T_G}{\Theta_1} + \frac{T_{si} - T_\infty}{\Theta_{conv}} = 1 \times 10^{-3}. \tag{H.15}$$

From equation (H.15) we have

$$T_{si} = \frac{\dfrac{T_G}{\Theta_1} + \dfrac{T_\infty}{\Theta_{conv}} + 1 \times 10^{-3}}{\dfrac{1}{\Theta_1} + \dfrac{1}{\Theta_{conv}}}. \tag{H.16}$$

Since

$$\Theta_1 = \frac{L}{kA} = \frac{1000 \times 10^{-6}}{133.5 \times 12 \times 10^{-12}} = 7.49 \times 10^5 \text{ K/W} \tag{H.17}$$

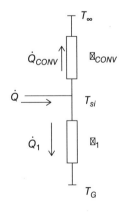

Figure H.4   Problem 8.6

and

$$\Theta_{\text{conv}} = \frac{1}{100 \times 4 \times 10^{-6}} = 2.5 \times 10^3, \tag{H.18}$$

we obtain

$$T_{si} = 302.5\,\text{K}. \tag{H.19}$$

Similarly, the power required to get this temperature up to 340 K is 16.04 mW.

Problem 8.7

In many thermal based MEMS devices, thermal isolation is needed around one area that should be at a higher temperature than its surroundings [62]. This can be achieved by interposing thermally isolating supports as illustrated in Problem 8.3, or by creating a bridge thinner than the supporting frame. This is illustrated in Figure H.5. Write the steady-state heat equation.

Compared to Problem 8.4, here the thermal conductivity is the same but the cross-section at the two edges of the bridge has a steep change as can be seen in Figure H.5. We have to change both the differential equation and the boundary conditions, as follows. Equation (H.10) applied to the bridge section becomes

$$k \left. \frac{d^2 T}{dx^2} \right|_x A_{\text{cond}} \Delta x = -\dot{Q}_{\text{gen}} = -\frac{P}{L} \Delta x, \tag{H.20}$$

where $A_{\text{cond}} = Wt$, and the boundary condition at $x = 0$ is a Neumann boundary condition. P is the power given to the device. Setting the heat rate equal on each side of $x = 0$,

$$kHW \left. \frac{dT}{dx} \right|_{x=0_-} = ktW \left. \frac{dT}{dx} \right|_{x=0_+}. \tag{H.21}$$

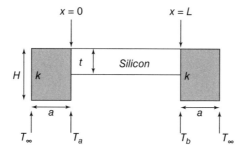

Figure H.5   Problem 8.7

To the left of $x = 0$, we have a conduction though a thick cross-section that can be written as

$$\left.\frac{dT}{dx}\right|_{x=0_-} = \frac{T_a - T_\infty}{a}, \tag{H.22}$$

and then

$$T_a = T_\infty + \frac{ta}{H}\left.\frac{dT}{dx}\right|_{x=0_+}. \tag{H.23}$$

Similarly, at the other edge of the bridge, $x = L$,

$$T_b = T_\infty - \frac{ta}{H}\left.\frac{dT}{dx}\right|_{x=L_-}. \tag{H.24}$$

Problem 8.8

For the same geometry as in Problem 8.7, assume that the power given to the device is $P = 10$ mW and that the thermal conductivity of the bridge and frame are the same $k = 133.5$ W/m·K. The bridge has thickness $t = 2\ \mu m$, length $L = 100\ \mu m$ and width $W = 10\ \mu m$, whereas the frame has thickness $H = 500\ \mu m$, length $a = 1000\ \mu m$ and width $W = 10\ \mu m$ equal than that of the bridge. Write Matlab code to solve for the temperature distribution along the bridge, and plot the result.

The main lines of the Matlab code are:

```
solinit= bvpinit(linspace(0,L,2000),[To 0])
sol=bvp4c(@problem8_6functionbook,@problem8_6bcbook,solinit);
x=linspace(0,L,500);
y=deval(sol,x);
```

where the function describing the differential equation is called,

```
function dydx=problem8_6functionbook(x,y)
dydx=[y(2)
   -P/(L*k*w*t)];
```

and the boundary conditions,

```
function res= problem8_6bcbook(ya,yb)
   res=[ya(1)-(To+t/H*a *ya(2))
   yb(1)-(To-t/H*a*yb(2))];
```

where 'To' is T_∞ in the equations above. The results are shown for the data values of this problem in Figure H.6.

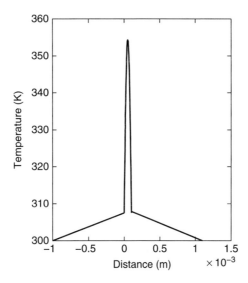

Figure H.6   Solution for Problem 8.8

Problem 8.9

Consider the geometry shown in Figure H.7, where there is a rod made of a material assumed to be a thermal insulator and covered with a thin film of a resistive material. Calculate the equations of the conduction and convection areas for a segment of the rod of length Δx. Write the equation of the differential resistance, dR, for this same segment. Write the steady-state heat equation.

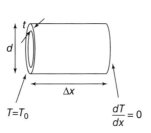

Figure H.7   Problem 8.9

The expressions for the convection and conduction areas are

$$A_{cond} = \pi\frac{d^2}{4} - \pi\frac{(d-t)^2}{4} = \frac{\pi}{4}(2dt - t^2), \quad A_{conv} = \pi d\Delta x, \tag{H.25}$$

and the generated heat rate

$$\dot{Q}_{gen} = I^2 dR. \tag{H.26}$$

The differential resistance is given by

$$dR = \rho\frac{\Delta x}{A_{cond}}. \tag{H.27}$$

Substituting in the heat equation,

$$\left.\frac{d^2T}{dx^2}\right|_x = \frac{h\pi d\Delta T}{kA_{cond}} - \frac{I^2\rho}{kA_{cond}^2}. \tag{H.28}$$

The boundary conditions are that at $x = 0$ the temperature is $T = T_0$, which is the temperature at the supports that we consider in equilibrium with the assembly of the wire, and we assume that the other end of the rod is a free end where the first derivative of the temperature is zero.

Problem 8.10

Write Matlab code to solve the steady-state heat flow equation derived in Problem 8.9 with a boundary condition at the position $x = L$, $dT/dx = 0$. Solve this boundary value problem for the following parameter values: $t = 0.635 \times 10^{-6}$, $k = 75$ W/m·K, $L = 1.02 \times 10^{-2}$ m, $I = 30$ mA and resistivity of the thin film material $\rho = 10.6 \times 10^{-8}$ Ω m.

We write the following code:

```
function res= viking1bcbook(ya,yb)
    res=[ya(1)-To
    yb(2)];
end
```

We also require the following solver:

```
function dydx=viking1functionbook(x,y)
Acond=pi/4*(2*d*t-t^2);
dydx=[y(2)
h*pi*d*(y(1)-Tair)/(k*Acond)-(I^2*resistivity)/(Acond^2*k)];
```

The solution is shown in Figure H.8. As can be seen, the temperature profile increases towards the free end of the rod, due to the boundary condition we have used.

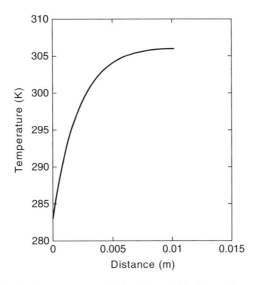

Figure H.8   Problem 8.10. Temperature profile in a heated thin film resistor on a rod with a free end

Problem 8.11

We have a flow meter in which a silicon volume of 1 *mm*^{2} *and* 500 *μm thick is heated by a platinum resistance. The chip is supported by two arms of length* 1 *mm, thermal conductivity* 1.005 *W/m·K and cross-section A* = 50 × 50 *μm*^{2}*. A flow is blown on the mass and the convection coefficient is h* = 200. *Write a PSpice code for the thermal circuit and solve the transient response, assuming that the power injected is* 10 *mW and that the fluid temperature is* 300 *K. The two supporting arms are connected to a PCB the temperature of which is assumed to be in equilibrium with the air,* T_{PCB} = 300 *K. The specific heat is c* = 1.005 *W/m·K.*

The equivalent thermal circuit is shown in Figure (H.9), where it can be seen that four circuit elements are involved: the thermal capacitance, C_{th}; the convection resistance, Θ_{conv}: and

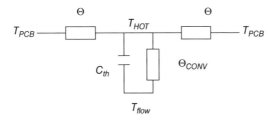

Figure H.9   Thermal equivalent circuit of Problem 8.11

the two thermal resistances Θ of the supports assumed equal in value. The values of these circuit elements are calculated from the parameter values and equations (H.16) and (8.51). The thermal resistance of the supports Θ is given by

$$\Theta = \frac{L}{kA} = \frac{1000 \times 10^{-6}}{1.005 \times 10 \times 10^{-6} \times 50 \times 10^{-6}} = 3.9 \times 10^{6} \, {}^\circ C/W, \tag{H.29}$$

and

$$C_{th} = c\,\rho_m vol = 703 \times 2340 \; 1000 \times 10^{-6} \times 1000 \times 10^{-6} \times 500 \times 10^{-6} = 8.22 \times 10^{-4}. \tag{H.30}$$

The PSpice code is as follows:

```
*thermal.cir
v1 1 0 0
v3 3 0 0
i1 0 2 50e-3
r1 1 2 3.9e6
r2 2 3 3.9e6
c 2 0 8.22e-4
vsense 30 0 0
eres 2 30 value= { i(vsense)* 1250*v(4,0)  }
```

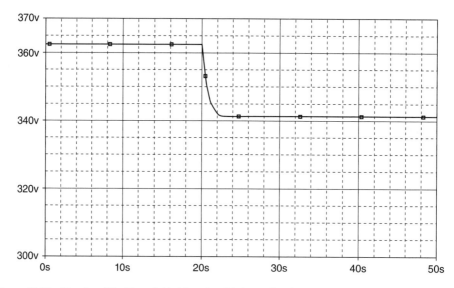

Figure H.10 Results of Problem 8.11. Note that this is an electrical equivalent of a thermal circuit and hence the units shown on the *y*-axis are volts, representing the equivalent value of temperature in kelvin.

```
vc 4 0 pulse ( 1 0.66 20 0.1 0.1 100 100)
.tran 1 100
.probe
.end
```

Here 'v1' and 'v3' are the temperature values at the PCB nodes, 'i1' is the current source representing the power injected, and 'r1' and 'r2' are the two Θ thermal resistors. 'vsense' is a zero-voltage source useful for measuring the current flowing through it ('i(vsense)') and serves to define a voltage-controlled resistor which is implemented by the voltage-controlled source 'eres'. The value assigned to 'eres' is the product of the current 'i(vsense)' and the convention resistor value. As we want in this example to show a transient, the convection resistance is assumed to take two values: 1250 at the beginning in steady state, with a sudden drop to two-thirds of this value at a given moment. This is implemented by the voltage control source 'vc' which is a voltage-pulsed source that takes two values: 1 at the beginning, and 2/3 after 20 seconds. The voltage of this source multiplies the product of the initial convection resistance by the value of 'i(vsense)'.

The ground node is assigned the value zero, and then the values of all nodes have to be reduced by $T_{flow} = 300$. The transient result is plotted in Figure H.10. As can be seen, initially the temperature of the hot point is 362.2 K; when the convection coefficient changes, the temperature drops to 341.2 K. Hence the temperature drops as the convection coefficient rises as expected. Similarly, the fall time of the dynamics of the temperature change can be estimated at 2 seconds.

I

Chapter 9 Solutions

Problem 9.1

Find an analytical model for dry oxidation of silicon and find the thickness of oxide that will grow at 1000°C in dry atmosphere in 1 hour.

The most popular model for silicon oxidation is found in [76]. For thin oxides we have the linear relationship

$$x = \frac{B}{A}t, \tag{I.1}$$

where B/A is the linear reaction constant which for dry oxidation is given by

$$\frac{B}{A} = \left(\frac{B}{A}\right)_0 e^{-E_A/kT} \tag{I.2}$$

with $(B/A)_0 = 3.71 \times 10^6$ μm/h for silicon (100), and $E_A = 2$ eV. Taking into account that $k = 8.617 \times 10^{-5}$ eV/K,

$$\frac{B}{A} = 3.71 \times 10^6 e^{-2/(8.617\times10^{-5}\times(1000+273))} = 44 \, \text{nm}. \tag{I.3}$$

Problem 9.2

We want to protect an area of bulk silicon oriented (100) against KOH. What should the thickness of oxide be in order to prevent etching of the protected area while removing 200 μm of the unprotected area? The concentration of KOH solution is 20% and the temperature 90°C.

Understanding MEMS: Principles and Applications, First Edition. Luis Castañer.
© 2016 John Wiley & Sons, Ltd. Published 2016 by John Wiley & Sons, Ltd.
Companion Website: www.wiley.com/go/castaner/understandingmems

The etch rates for bulk silicon (100) and silicon dioxide given in [77] are $ER_{Si} = 150\,\mu m/h$ and $ER_{SiO_2} = 580\,nm/h$, respectively. The resulting etch rate ratio (ERR) is

$$ERR = \frac{ER_{Si}}{ER_{SiO_2}} = \frac{150 \times 10^{-6}}{580 \times 10^{-9}} = 258.6. \qquad (I.4)$$

If the silicon etchant etches away $200\,\mu m$, the resulting thickness etched on the oxide is

$$\frac{t_{Si}}{t_{SiO_2}} = ERR. \qquad (I.5)$$

Then

$$t_{SiO_2} = \frac{t_{Si}}{ERR} = \frac{200 \times 10^{-6}}{258.6} = 0.77\,\mu m. \qquad (I.6)$$

At least $0.77\,\mu m$ of silicon dioxide must be on top of the area to be protected.

Problem 9.3

In Example 9.7.1 an isotropic etch is performed to totally undercut the silicon below the bimorph area. Calculate the minimum time required if a hydrofluoric–nitric–acetic acid mixture etching solution is used.

In [78], for a hydrofluoric–nitric–acetic acid solution 3 : 2 : 5 by volume, we find an etch rate value of $\sim 20\,\mu m/min$. The bimorph is $10\,\mu m$ wide and the single-crystal silicon membrane where the bimorph has to be located is $45\,\mu m$ thick. As the etching will proceed 9n all directions, full etch will first be achieved underneath the bimorph in the lateral direction. As the width is $10\,\mu m$, the answer to the problem will be the time required to etch $5\,\mu m$ from each side. At the rate of $20\,\mu m/min$, the time required is 15 seconds. The walls of the membrane and underneath the mirror will also be undercut by $5\,\mu m$ each side.

Problem 9.4

In Example 9.6.1, for the fabrication of a cantilever, silicon dioxide is used as a sacrificial layer. The release of the cantilever requires the silicon dioxide to be etched selectively. From Figure 9.7, identify the materials that are in contact with the silicon dioxide before the release step and assess the selectivity of the silicon dioxide etch.

Looking into the materials that are in contact with the silicon dioxide we find:

- structural layer (amorphous silicon phosphorus doped);
- dielectric layer (hafnium oxide);
- insulating layer (silicon nitride);
- metal (chromium).

We can find in the literature the common etchants for the materials listed:

- amorphous silicon: KOH, isopropyl alcohol (IPA) and H_2O [79];
- hafnium oxide: HF/deionized water and HF/IPA [80];
- silicon nitride etchant: H_3PO_4 (85%, aq) [81];
- chromium etchant: 2 $KMnO_4$(s) : 3 NaOH(s) : 12 H_2O [81].

Hydrofluoric acid etches the following materials: GaAs, Ni, SiO_2, Ti and Al_2O_3. As can be seen, the only problem may be the selectivity of the etch between hafnium oxide and silicon dioxide. In reference [80], with sufficient dilution of the HF in water or IPA sufficient selectivity is reached.

Problem 9.5

Calculate the concentration of impurities at a depth of 0.5 μm for drive-in times of 15 minutes if the total amount of impurities implanted is 5×10^{18} cm^{-2}. The drive-in temperature is $1050°C$ and the diffusivity is $D = 9.3 \times 10^{-14}$ cm^2/s.

Using the concentration distribution equation (I.1),

$$C(x) = \frac{S}{\sqrt{\pi D t}} e^{-x^2/4Dt},$$

for the same diffusivity values and for 15 minutes, we have $C = 1.75 \times 10^{20} cm^3$.

Problem 9.6

For a diffusion of impurities into a bulk material from an unlimited source, the depth follows a complementary error function distribution. Calculate the total amount of boron impurities diffused into silicon after 30 minutes' diffusion at $900°C$.

The complementary error function distribution, denoted erfc, is given by

$$N(x,t) = N_s \, \mathrm{erfc}\left(\frac{x}{2\sqrt{Dt}}\right), \tag{I.7}$$

where N_s is the concentration at the surface, D the diffusivity and t time. The solid solubility of boron at $900°C$ is 3.9×10^{20} cm^{-3} and the diffusivity is $D = 6.6 \times 10^{-15}$ cm^2/s. So we can assume that the surface concentration is equal to the solid solubility. The amount of impurities is given by

$$S = \int_0^\infty N_s \, \mathrm{erfc}\left(\frac{x}{2\sqrt{Dt}}\right) dx. \tag{I.8}$$

From [82], we have

$$\int_0^\infty \text{erfc}(ax)dx = \frac{1}{a\sqrt{\pi}}. \tag{I.9}$$

Hence,

$$S = \int_0^\infty N_s\text{erfc}\left(\frac{x}{2\sqrt{Dt}}\right)dx = 2N_s\sqrt{\frac{Dt}{\pi}} = 2.95 \times 10^9 \text{ cm}^{-2}. \tag{I.10}$$

Problem 9.7

For the same data as in Problem 9.6, calculate the junction depth if such diffusion is carried out on silicon doped with $N_B = 10^{17}$ cm^{-3} n-type.

The junction depth is defined as the depth inside the substrate where the boron concentration becomes equal to the substrate concentration, thus

$$N(x_j, t) = N_s \, \text{erfc}(x_j/2\sqrt{Dt}) = N_B \tag{I.11}$$

and

$$x_j = 2\sqrt{Dt}\,\text{erfc}^{-1}\left(\frac{N_B}{N_s}\right) = 1.78 \times 10^{-5} \text{ cm}. \tag{I.12}$$

This can be seen in Figure I.1.

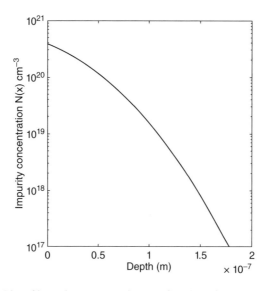

Figure I.1   Plot of impurity concentration as a function of depth for Problem 9.7

Problem 9.8

Using the same data as in Problem 9.6, calculate the value of the sheet resistance resulting after the diffusion process described in Problem 9.6. Assume that the hole mobility is given by

$$\mu = \mu_{min} + \frac{\mu_{max} - \mu_{min}}{1 + (N(x)/N_r)^\alpha}.$$ (I.13)

For boron diffusion into silicon, the values of the constants in equation (I.13) are $\mu_{min} =$ 44.9 cm^2/Vs, $\mu_{max} = 470.5$ cm^2/Vs, $N_r = 2.23 \times 10^{17}$ cm^{-3} and $\alpha = 0.719$.

In a diffused layer the resistivity is a function of the depth, and hence the sheet resistance, defined as

$$R^\square = \frac{\rho}{t},$$ (I.14)

for a layer of constant resistivity value and thickness t, can be rewritten for a diffused layer as

$$R^\square = \frac{\overline{\rho}}{x_j},$$ (I.15)

where $\overline{\rho}$ is the average resistivity value in the layer and we have set $t = x_j$ as a diffused layer is made on a substrate of opposite doping and hence the thickness of the diffused layer is the junction depth x_j. Equation (I.15) can be written as

$$R^\square = \frac{1}{x_j\overline{\sigma}} = \frac{1}{x_j \frac{1}{x_j} \int_0^{x_j} q\mu N(x)dx},$$ (I.16)

where $\mu(x)$ is the mobility of majority carriers (holes). The distribution of impurities after diffusion can be approximated by a Gaussian function,

$$N(x) = \frac{S}{\sqrt{\pi Dt}} e^{-x^2/4Dt}.$$ (I.17)

and the mobility is given by equation (I.13).
   Equation (I.16) gives a value for the sheet resistance of 82.6 $\Omega/\square$.

Problem 9.9

In [71] a membrane strain sensor is described where n-doped polysilicon thin films are used as piezoresistances, on top of membrane made of a stack of silicon dioxide/silicon nitride layers. Draw the process flow and estimate the resistivity of the polysilicon layer to achieve 4.6 $k\Omega$ for a width $W = 20$ μm. The polysilicon strip is $t = 0.35$ μm thick and $L = 800$ μm long.

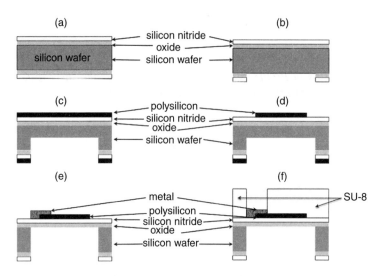

Figure I.2 Problem 9.9. Process flow adapted from [71]: (a) thermal oxidation of wafer to grow silicon dioxide followed by LPCVD deposition of silicon nitride; (b) back side patterning of silicon dioxide and silicon nitride to open window for membrane etching; (c) LPCVD n-type doped polysilicon deposition, back side DRIE polysilicon patterning and partial membrane etch; (d) front side polysilicon patterning to form the resistors; (e) front metal deposition and patterning; (f) back side DRIE etch of membrane until stop at the silicon dioxide layer, deposition of SU-8 and patterning

The process flow is illustrated in Figure I.2, where it can be seen that the membrane is composed of two layers, one of silicon dioxide and the other of silicon nitride. The membrane is formed by using these two layers as a mask for the DRIE silicon etch. SU-8 polymer is used to protect the thin membrane and the thin polysilicon patterns.

As the width of the polysilicon pattern is 20 μm, in a length of 800 μm we have $800/20 = 40$ squares. By the definition of a resistance,

$$R = R^\square \frac{L}{W} = R^\square \frac{800}{20} = 40R^\square. \tag{I.18}$$

Hence,

$$R^\square = 115\,\Omega/\square. \tag{I.19}$$

As the thickness of the polysilicon layer is, according to [71], 0.3–0.4 μm, taking $t = 0.35$ μm, the resistivity is

$$\rho = R^\square t = 115\Omega/\square \times 0.35 \times 10^{-6} m = 4.02 \times 10^{-5}\,\Omega\,m. \tag{I.20}$$

Problem 9.10

In [72] the piezoresistive and thermoelectric effects of carbon nanotubes are measured. To do so, a silicon nitride film is deposited on a bare silicon wafer. A layer of chromium/gold is patterned to provide measurement electrodes. A forest film of single-wall nanotubes is then deposited, partially covering the electrodes. Draw the process flow schematically.

The process is shown in Figure I.3, adapted from [72].

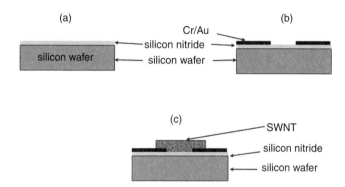

Figure I.3 Problem 9.10: (a) silicon nitride film deposited on a bare silicon wafer; (b) a stack of chromium/gold layers is patterned by lift-off on top; (c) single-wall nanotubes previously synthesized are deposited, covering part of the electrodes and then patterned.

Problem 9.11

In [40] a resonating gravimetric sensor fabrication process is described. The main properties of this device were analysed in Problem 6.4. Draw the process flow as a combination of silicon on glass and silicon on insulator techniques.

The fabrication of this sensor is based on two wafers: one of SOI and the second of glass. The SOI wafer is used to pattern the proof mass, the flexures and the comb drive actuator. This is done by standard patterning and DRIE of the active device layer of the SOI wafer. The buried oxide serves as an etch stop. The glass wafer is processed independently by depositing and patterning a Cr/Au layer that serves as a hard mask for the HF etching of the glass. This step creates trenches in the glass wafer surface. Once this is done, the SOI wafer is flipped on the glass wafer and the two are anodically bonded. Finally, the handle wafer of the SOI wafer is dissolved and the buried oxide etched, leaving the proof mass free to resonate in the cavity created by the trench in the glass wafer. This is shown in Figure I.4.

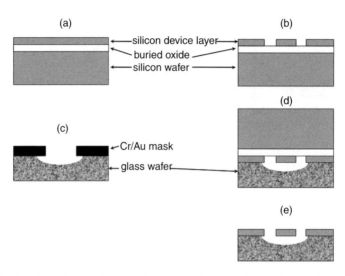

Figure I.4 Problem 9.11. Process flow adapted from [40]: (a) starting SOI wafer; (b) SOI wafer device layer patterned to create proof mass, flexures and comb drive; (c) glass wafer front etched isotropically using a Cr/Au hard mask; (d) flip-chip SOI wafer and anodic silicon–glass bonding; (e) SOI handle wafer is dissolved and buried oxide etched.

References

[1] D. J. Kleppner and R. J. Kolenkow, *An Introduction to Mechanics*, McGraw-Hill (1973), p. 79.

[2] S. Ramo, J. R. Whinnery and T. Van Duzer, *Fields and Waves in Communication Electronics*, Wiley (1965), p. 70.

[3] P.-G. de Gennes, F. Brochard-Wyart and D. Quéré, *Capillarity and Wetting Phenomena: Drops, Bubbles, Pearls, Waves*, Springer (2003), p. 3.

[4] J.-M. Cho, S.-Y. Lee, S.-W. Kim, K.-S. Kim and S. An, A Lorentz force based fusion magnetometer-accelerometer with dual functions for the electronic compass, *Applied Physics Letters*, 91, 203519 (2007).

[5] G. Langfelder, G. Laghi, P. Minotti, A. Tocchio and A. Longoni, Off-resonance low-pressure operation of Lorentz force MEMS magnetometers, *IEEE Transactions on Industrial Electronics*, 61, 7124–7130 (2014).

[6] T. B. Jones, *Electromechanics of Particles*, Cambridge University Press (1995), p. 5.

[7] S. H. Crandall, N. C. Dahl and T. J. Lardner, *An Introduction to the Mechanics of Solids*, 2nd edn, McGraw-Hill (1978), p. 201.

[8] M. A. Hopcroft, W. D. Nix and T. W. Kenny, What is the Young's modulus of silicon? *Journal of Microelectromechanical Systems*, 19, 229–238 (2010).

[9] E. Kreyszig, *Advanced Engineering Mathematics*, 3rd edn, Wiley (1962), p. 281.

[10] W. C. Young, R. G. Budynas, *Roark's Formulas for Stress and Strain*, McGraw-Hill (2002).

[11] W. P. Mason, *Physical Acoustics and the Properties of Solids*, Van Nostrand (1958), p. 355.

[12] J. J. Wortman and R. A. Evans, Young's modulus, shear modulus and Poisson's ratio in silicon and germanium, *Journal of Applied Physics*, 36, 153–156 (1965)

[13] http://www.azom.com

[14] W. N. Sharpe, B. Yuan, R. L. Edwards and R. Vaidyanathan, Measurements of Young's modulus, Poisson's ratio, and tensile strength of polysilicon, *Proceedings of the 10th International Workshop on Microelectromecanical Systems*, Nagoya, (1997), pp. 424–429.

[15] S. D. Senturia, *Microsystem Design*, Kluwer Academic (2001).

[16] G. K. Fedder, System level simulations of microsystems, in J. G. Korvink and O. Paul (eds), *MEMS: A Practical Guide to Design Analysis and Applications*, William Andrew Publishing (2006).

[17] S. Timoshenko and S. Woinowsky-Krieger, *Theory of Plates and Shells*, McGraw-Hill (1959), p. 82.

[18] S. K. Clark and K. D. Wise, Pressure sensitivity in anisotropically etched thin-diaphragm pressure sensors, *IEEE Transactions on Electron Devices*, ED-26, 1887–1896 (1979).

[19] M.-H. Bao, *Micro Mechanical Transducers: Pressure Sensors, Accelerometers and Gyroscopes*, Elsevier (2000).

[20] E. J. Hearn, *Mechanics of Materials 2*, 3rd edn, Butterworth-Heinemann (1997) p. 195.

[21] J. T. Riet, A. J. Katan, C. Rankl, S. W. Stahl, A. M. van Buul, I. Y. Phang, A. Gomez-Casado, P. Schon, J. W. Gerritsen, A. Cambi, A. E. Rowan, G. J. Vancso, P. Jonkheim, J. Huskens, T. H. Oosterkamp, H. Gaub, P. Hinterdorfer, C. G. Figdor, S. Speller, Interlaboratory round robin on cantilever calibration for AFM force spectroscopy, *Ultramicroscopy* 111, 1659–1669 (2011).

[22] F. M. Smits, Measurement of sheet resistivities with the four-point probe. *Bell System Technical Journal*, 34, 711–718 (1958).

Understanding MEMS: Principles and Applications, First Edition. Luis Castañer.
© 2016 John Wiley & Sons, Ltd. Published 2016 by John Wiley & Sons, Ltd.
Companion Website: www.wiley.com/go/castaner/understandingmems

[23] W. P. Mason and R. N. Thurston, Use of piezoresistive materials in the measurement of displacement, force, and torque, *Journal of the Acoustical Society of America*, 29, 1096–1101, (1957).

[24] C. S. Smith, Piezoresistance effect in silicon and germanium, *Physical Review*, 94, 42–49 (1954).

[25] H. P. Phan, D. V. Dao, P. Tanner, L. Wang, N. T. Nguyen, Y. Zhu, S. Dimitrijev, Fundamental piezoresistive coefficients of p-type single crystalline 3C-SiC, *Applied Physics Letters*, 104, 111905 (2014).

[26] G. K. Johns, Modeling piezoresistivity in silicon and polysilicon, *Journal of Applied Engineering Mathematics*, 2, 1–5 (2006).

[27] http://www.efunda.com/materials

[28] F. Gerfers, P. M. Kohlstadt, E. Ginsburg, M. Yuan He, D. Samara-Rubio, Y. Manoli and L.-P. Wang, Sputtered AlN thin films for piezoelectric MEMS devices – FBAR resonators and accelerometers, in J. W. Swart (ed.), *Solid State Circuits Technologies*, InTech (2010).

[29] G. Piazza, Piezoelectric aluminum nitride vibrating RF MEMS for radio front end technology, PhD thesis, University of California Berkeley (2005).

[30] S. H. Crandall, D. C. Karnopp, D. C. Pridmore-Brown, *Dynamics of Mechanical and Electromechanical Systems*, Krieger (1968), p. 260.

[31] H. C. Nathanson, W. E. Nevel, R. A. Wickstrom, and J. R. Davis Jr., The resonant gate transistor, *IEEE Transactions on Electron Devices*, ED-14, 117–133 (1967).

[32] P. G. Steeneken and O. Wunnicke, Performance limits of MEMS switches for power electronics, *Proceedings of the 24th International Symposium on Power Semiconductor Devices and ICs*, IEEE (2012), pp. 417–420.

[33] L. Castañer and S. D. Senturia, Speed-energy optimization of electrostatic actuators based on pull-in, *IEEE Journal of Microelectromechanical Systems*, 8, 290–298 (1999)

[34] L. Castañer, J. Pons, R. Nadal-Guardia and A. Rodriguez, Analysis of the extended operation range of electrostatic actuators by current-pulse drive, *Sensors and Actuators A*, 90, 181–190 (2001).

[35] ADXL50 Monolithic Accelerometer with Signal Conditioning, Product Data Sheet, Analog Devices (1995).

[36] R. Sattler, F. Plotz, G. Fattinger and G. Wachutka, Modeling of an electrostatic torsional actuator: demonstrated with an RF MEMS switch, *Sensors and Actuators A*, 97–98, 337–346 (2002).

[37] G. N. Nielson and G. Barbastathis, Dynamic pull-in of parallel plate and torsional electrostatic MEMS actuators, *IEEE Journal of Microelectromechanical Systems*, 15, 811–821 (2006)

[38] W. T. Thomson, *Theory of Vibration with Applications*, 4th edn, Prentice Hall (1993), p. 384.

[39] R. Wang, S. K. Durgam, Z. Hao and L. L. Vahala, An SOI-based tuning-fork gyroscope with high quality factors, *Proc. SPIE*, 7292, Sensors and Smart Structures Technologies for Civil, Mechanical, and Aerospace Systems 2009, 729238 (30 March).

[40] D. Eroglu, E. Bayraktar and H. Kulah, A laterally resonating gravimetric sensor with uniform mass sensitivity and high linearity, *16th International Solid-State Sensors, Actuators and Microsystems Conference* (TRANSDUCERS), Beijing, pp. 2255-2258 (2011).

[41] S. Bouwstra, B. Geijselaers, On the resonance frequencies of microbridges, *1991 International Conference on Solid-State Sensors, Actuators and Microsystems*, pp. 538–542 (1991).

[42] J. Li, S. Fan, Z. Guo, Design and analysis of silicon resonant accelerometer, *Research Journal of Applied Sciences, Engineering and Technology*, 5, 970–974 (2013).

[43] J. A. Fay, *Introduction to Fluid Mechanics*, MIT Press (1994), p. 265.

[44] A. W. Adamson and A. P. Gast, *Physical Chemistry of Surfaces*, 6th edn, Wiley (1997), p. 169.

[45] F. Mugele and J.C. Baret, Electrowetting: from basics to applications, *Journal of Physics: Condensed Matter*, 17(28), R705 (2005).

[46] C. Quilliet and B. Berge, Electrowetting: a recent outbreak, *Current Opinion in Colloid & Interface Science*, 6, 34–39 (2001).

[47] C.-X. Liu, J. Park, and J.-W. Choi, A planar lens based on electrowetting of two immiscible liquids, *Journal of Micromechanics and Microengineering*, 18, 035023 (2008).

[48] F. Ouyang, J. Wu, M. Kang, R. Yue, and L. Liu, Planar variable-focus liquid lens based on electrowetting on dielectric, *Conference on Nano/Micro Engineered and Molecular Systems*, Bangkok (2007).

[49] L. Castañer, V. Di Virgilio and S. Bermejo, Charge-coupled transient model for electrowetting, *Langmuir*, 26, 16178–16185 (2010).

[50] G. H. Markx, R. Pethig and J. Rousselet, The dielctrophoretic levitation of latex beads with reference to field-flow fractionation, *Journal of Physics D: Applied Physics*, 30, 2470–2477 (1997).

[51] A. Bejan, *Heat Transfer*, Wiley (1993), p. 7.

[52] http://www.km.kongsberg.com/ks/web/nokbg0397.nsf/AllWeb/A707D00EE0F558D6C12574E1002C2D1C/$file/tsiec751_ce.pdf

[53] M. Domínguez, V. Jiménez, J. Ricart, L. Kowalski, J: Torres, S. Navarro, J. Romeral and L. Castañer, A hot film anemometer for the Martian atmosphere, *Planetary and Space Science*, 56, 1169–1179 (2008).

[54] S. Silvestri and E. Schena, Micromachined flow sensors in biomedical applications, *Micromachines*, 3, 225–243, 2012.

[55] X. Zhang, H. Xie, M. Fujii, H. Ago, K. Takahashi, T. Ikuta, H. Abe and T. Shimizu, Thermal and electrical conductivity of a suspended platinum nanofilm, *Applied Physics Letters*, 86, 171912 (2005)

[56] F. P. Incropera and D. P. DeWitt, *Fundamentals of Heat and Mass Transfer*, 5th edition, Wiley (2002), p. 436.

[57] H. Y. Wu and P. Cheng, An experimental study of convective heat transfer in silicon microchannels with different surface conditions, *International Journal of Heat and Mass Transfer*, 46, 2547–2556 (2003).

[58] Y. Ito, T. Higuchi and K. Takahashi, Submicroscale flow sensor employing suspended hot film with carbon nanotube fins, *Journal of Thermal Science and Technology*, 5, 51–60 (2010)

[59] S. Timoshenko, Analysis of bi-metal thermostats, *Journal of the Optical Society of America*, 11, 233–255 (1925).

[60] A. Jain, H. Qu, S. Todd, H. Xie, A thermal bimorph micromirror with large bi-directional and vertical actuation, *Sensors and Actuators A*, 122, 9–15 (2005).

[61] C. G. Matsson, G. Thungstöm, K. Bertilsson, H.-E. Nilsson and H. Martin, Fabrication and characterization of a design optimized SU-8 thermopile with enhanced sensitivity, *Measurement Science and Technology*, 20, 115202 (2009).

[62] D. Randjelovic, A. Petropoulos, G. Kaltsas, M. Stojanovic, Z. Lazic, Z. Djuric and M. Matic, Multipurpose MEMS thermal sensor based on thermopiles, *Sensors and Actiuators A*, 141, 403–4013 (2008).

[63] M. Rosa, Thick epitaxial silicon enhances performance of power and MEMS devices, *Nanofab*, 9, 24–32 (2014).

[64] M. Elwenspoek and H. Jansen, *Silicon Micromachining*, Cambridge University Press (1998).

[65] Y. Qing Fu, Jack Luo, S. Milne, A. Flewitt, Process flow simulation and manufacturing of variable RF MEMS capacitors, www.silvaco.com (Silvaco Technical library, pp. 5–8 (August 2005).

[66] Method of anisotropic etching of silicon, United States Patent 6531068.

[67] H. Xie, L. Erdmann, X. Zhu, K. J. Gabiel and G. K. Fedder, Post-CMOS processing for high aspect ratio integrated silicon microstructures, *Journal of Microelectromechanical Systems*, 11, 93–101 (2002).

[68] P. Ortega, S. Bermejo and L. Castañer, High voltage photovoltaic minimodules, *Progress in Photovoltaics: Research and Applications*, 16, 369–377 (2008).

[69] A Cowen, B. Hardy, R. Mahadevan and S. Wilcenski, *PolyMUMPs Design Handbook*, http://www.memscap.com/_data/assets/pdf_file/0019/1729/PolyMUMPs-DR-13-0.pdf

[70] S. Haghighat, Design and analysis of a MEMS Fabry-Perot pressure sensor, MS thesis, University of Waterloo (2007).

[71] L. Cao, T. S. Kimm, S. C. Mantell and D. L. Polla, Simulation and fabrication of piezoresistive membrane type MEMS strain sensors, *Sensors and Actuators*, 80, 273–279 (2000).

[72] V. T. Dau, T. Yamada, D. Z. Dao, B. T. Tung, K. Hata and S. Sugiyama, Piezoresistive and thermoelectric effects of CNT thin film patterned by EB lithography, *IEEE Sensors Conference*, pp. 1048–1051 (2009).

[73] G. K. Fedder, Simulation of micromechanical systems, PhD thesis, University of California Berkeley (1994).

[74] F. B. Beer, E. R. Johnston Jr. and J. T. De Wolf, *Mechanics of Materials*, 3rd edn, McGraw-Hill (2004), p. 736.

[75] A. Rampal, Experimental and theoretical study of quartz and InP piezoelectric MEMS resonators, MS thesis, McMaster University (2010).

[76] B. E. Deal and A. S. Grove, General relationship for the thermal oxidation of silicon, *Journal of Applied Physics*, 36, 3770–3778 (1965).

[77] H. Seidel, L. Csepregi, A. Heuberger and H. Baumgärtel, Anisotropic etching of crystalline silicon in alkaline solutions: I. Orientation dependence and behavior of passivation layers. *Journal of the Electrochemical Society*, 137, 3612–3626 (1990).

[78] https://www.ee.washington.edu/research/microtech/cam/PROCESSES/PDF%20FILES/WetEtching.pdf

[79] Y. S. Tsuoa, Y. Xua, D. W. Bakera and S.K Deba, Etching properties of hydrogenated amorphous silicon, *MRS Proceedings*, 219. 805–810 (1991).

[80] T.-K. Kang, C.-C. Wang, B.-Y. Tsui, Y.-H. Li, Selectivity investigation of HfO_2 to oxide using wet etching, *Semiconductor Manufacturing Technology Workshop Proceedings*, IEEE (2004), pp. 87–90.

[81] http://cleanroom.ien.gatech.edu/media/resources/processing/metal_etchants.pdf

[82] E. W. Ng and M. Geller, A table of integrals of the error functions, *Journal of Research of the National Bureau of Standards B: Mathematical Sciences*, 73B(1) (1969).

Index

Understanding MEMS: Principles and Applications, First Edition. Luis Castañer.
© 2016 John Wiley & Sons, Ltd. Published 2016 by John Wiley & Sons, Ltd.
Companion Website: www.wiley.com/go/castaner/understandingmems